MW01590857

Time Just Flew

Time Just Flew

To John
A good friend.
Jim McCorkle

Jim McCorkle

and

Lorraine (McCorkle) Hart

Key Literary Concepts

Printed in the United States of America

First Edition: February 2015

McCorkle, Jim

Library of Congress Control Number 2015934025

ISBN **978-1508501367**

10 9 8 7 6 5 4 3 2 1

Dedication

These stories are dedicated to the good men and women who served with me in WWII, especially Corporal Bunnie Keelan, who stood by my side for the rest of her life.

I'll see you when I get back.

INTRODUCTION

My father treated every fixed-wing and helicopter he flew like it was a living creature, and each machine responded to his skill as if it were true. During WWII he earned the nickname "Lucky" and a few badges for always making it back, when so many didn't. He never spoke much about his wartime experiences when I was younger, preferring to tell stories about all the places he had seen, and the many cultures he had experienced. We are forever indebted to his generation for the harrowing work they performed so heroically.

It has been my honour to help my dad bring just some of his lifetime stories together for this book. Working on them has enabled me to see him from a new perspective with each story, gaining a new respect for the man who logged over fifty-thousand hours in the cockpit before retiring. Some of our happiest times together were spent in helicopters and hangars. At ninety-three he can still weave a tale, still fly a dream.

TIME

Time is elastic, stretching endlessly,

Encompassing everything around.

Sometimes it flies on golden wings,

But at others, runs aground.

To impatient youth, it seems eternity

Between each passing day,

Yet inexorably it ticks the hours away,

Unchanging in its tempo.

Hold fast, stand still, we sometimes want to cry,

I need to hold this moment, never let it fly,

But through our hands the sands of time just stray

Like quicksilver drops in the path of destiny.

A precious thing, a golden thing, one not to be abused,

So take each fleeting second, make sure it is well used,

Waste not a single moment.

This thought is quite profound,

Remember simply this, my friends,

TIME LOST IS NEVER FOUND.

MY FIRST SOLO FLIGHT

In the life of every pilot, young or old, male or female, there is always one special event which will be remembered in vivid, startling detail, and almost total recall. I refer of course to one's first solo flight. For me, this momentous occasion took place at a small grass airfield on the outskirts of Peterborough, in England. Known as #17 E.F.T.S. (Elementary Flying Training School) of the Royal Air Force, where I was first introduced to the thrilling experience of powered flight. The aircraft in vogue at the time, as a trainer, was the De Havilland "Tiger Moth" known affectionately by us wartime students as the "Tigerschmidt" a sort of back-handed reference to the German aircraft we all expected to confront at some vague time in our near future.

The year was 1941. On a beautiful Spring morning I was airborne for the third time in a Moth with my instructor, Flying Officer Barclay, performing what we called "Circuits and Bumps," taking off into the wind, executing a left-handed traffic pattern involving a climb to 1,000 feet above ground level, and landing into the wind on the grass. All went well as far as I could judge and, at the end of the detail with a total of four flying hours to my credit, I taxied the yellow aeroplane to a point close to the dispersal area. With a curt, "Keep this motor ticking over," my instructor clambered from the front cockpit, released his parachute harness and slung it over his left shoulder, striding away to disappear inside the Flight Commander's office just a few yards away.

I didn't have a clue about what was happening and, when I saw Flight Lieutenant Scott, Chief Flying Instructor, emerge from the shack my first thought was, God he's going to scrub me! Moments later "Scotty" as he was fondly known, lowered himself into the cockpit, secured his safety harness, and connected the Gosport Tube, which allowed voice communication between instructor and student; no radio in these early training machines.

"Right young McCorkle," he said, "Take me out and do a circuit of the field."

Still unsure of why I had been singled out to fly with the head man, I slowly trundled the Tiger Moth out toward the take-off area. The slightly long grass and the metal tail skid (no wheel) made taxiing out quite a chore, involving heavy use of the rudder pedals, and bursts of throttle to swing the tail around. Finally, parked across wind at ninety degrees to the landing and take-off run, I completed the pre-take-off checks and eyeballed the final approach, to ensure my departure run would not endanger anyone about to land.

I turned into the wind and performed the take-off run fairly mechanically, still semi-convinced that this check ride had ominous overtones for me. I strove to make the 1,000 foot circuit as accurate as possible, and the turns properly balanced without slip or skid; all this time, not a single word from the front cockpit, neither praise nor condemnation. Despite the cool air on my face, I was perspiring profusely inside my bulky flying suit, and my knuckles were white on the controls. The landing was achieved with only a minor ballooning during the round-out, and a fairly presentable three-point touchdown, much to my relief.

As we taxied in, Flt.Lt. Scott suddenly pulled the throttle closed and ordered me to hold our present position across wind. Next thing I knew, he was climbing out of his cockpit and walking around the wingtip of our machine, his 'chute banging clumsily against the back of his legs. I watched in wonder as he whipped off his leather helmet. He then stooped close to the ground and proceeded to pluck something from the grass. This performance was repeated several times before he returned to the side of the aircraft. Leaning in from the noisy slipstream he yelled to me.

"Do you like mushrooms?"

"I don't know Sir," I replied, "I've never tasted them."

Opening a tiny baggage compartment, just aft of my back rest, he deposited the contents of his now full helmet therein. He then removed his parachute, and leaned into the empty front cockpit to secure the loosened four-point safety harness, preventing it from getting in the way of any controls.

With a nonchalant, "Take her up on your own, and do a couple of circuits and landings," he tapped me on the left shoulder and strolled away without looking back.

For a few moments the portent of what he had said did not strike home, and I gaped at his retreating figure as he hauled his 'chute onto his shoulder. I stared at the empty cockpit ahead of me, usually filled with the bulk of the instructor's head and shoulders, and then the adrenalin rush hit me. I was alone in an aeroplane, about to become airborne with myself as the sole occupant. No radio. No air-traffic controller handy to offer advice. I was about to fly solo for the very first time.

Parked across-wind as I was, all I had to do was check for other aircraft movements, and get my act on the road or rather, into the air. I felt a terrific sense of elation mixed with just a tiny twinge of uncertainty as I swung the nose around to point toward the upwind end of the 'drome. The hum of the Gypsy Major engine never sounded so sweet before, and the blur of the small wooden propeller was clearly visible as I opened the throttle. All of my senses seemed more acute, and the control column (joystick) in my right hand transmitted every vibration, every bump, to fingers which had never felt more alive.

Long before the end of the field drew close, I exerted a gentle back pressure on the stick, feeling that the Moth was ready and willing to fly. The transition from being an earth-borne creature to one in the air was discerned by a change in the vibration level, and a swooping gain in height. "Watch that nose attitude," I could hear my invisible instructor admonish, just as if his bulk was still present in the front seat. I automatically adjusted the elevator control to bring the nose cowling down to its normal outside reference point of the natural horizon.

The sky was a deep blue, interspersed with miniature patches of fair weather cumulus clouds, looking like haphazardly tossed puffs of cotton wool. Had I known how to perform aerobatics at this time, there's nothing in this world could have prevented me from executing a series of loops and rolls. I was so pumped-up with exhilaration that I just had to give vent to my feelings. I yelled out loud the words to a popular song of the day, with the slipstream buffeting against my goggled eyes.

"He flies through the air with the greatest of ease, that daring young man on the flying trapeze!" The words were whipped away, heard by no one but myself. I suppose this phenomenon could be likened to a diver's experience known as Rhapsody of the Deep.

At 500 feet above ground level the trees and green fields stood out as clearly as if etched in crystal. One detached portion of my mind registered all ground movement of personnel and vehicles around the military camp, while another performed the function of entering a climbing left-hand turn to crosswind, making allowance for wind drift. As the altimeter (one of the few instruments on the meagre panel) registered 1,000 feet I leveled-out and turned downwind. A gentle dip of the left wing served to check that I was correctly spaced outboard of the field, with the wingtip running along the country lane which bordered the camp. My still absent tutor took me through the routine downwind pre-landing checks, and I listened to the airstream rushing past the struts, separating and supporting the upper and lower fabric-covered wings.

The airspeed indicator in these early Tiger Moths was a Heath Robinson affair. Attached to the nearest of the vertical struts on the left wing was a metal flange, with a pointed end. This was deflected rearward by the action of rushing air, and indicated increments of airspeed against an engraved metal plate. It was primitive, but quite effective for this low-powered training aircraft. I can remember one instructor saying, with a straight face,

"If you can hear the wind whistling between the bracing wires, you are going too fast, and if you can't hear the wind, you're going too slow and about to stall," very reassuring to a novice flyer!

Once settled on the so-called base leg of the circuit, I eased the throttle back to idle and commenced a glide approach, giving the engine occasional bursts of power to avoid plug-fouling, or carb-icing.

Rather than feeling I was descending, I felt that the airfield was floating upward toward me and again, in my heightened awareness, every detail below me seemed startlingly clear. Over the hedge at just the right height and speed, I closed the throttle completely, and eased the stick back ever so gently. The lovely old Moth settled smoothly onto the grass, in a perfect three-pointer…the best I had accomplished to date. Hope my instructor was watching that, I thought fleetingly, as I opened the throttle and commenced my second takeoff.

The second circuit followed exactly as the first, with my feeling of exultation in no way diminished. I felt as if some master hand was guiding the controls, and even the twinge of apprehension had disappeared from the edge of my consciousness. It was bliss, pure and simple. I now seemed to have more time to look around me and take in all the sights. An old recruiting slogan of the Royal Air Force popped into my mind with new meaning, "See the world from a new angle!" The sense of remoteness, freedom, and detachment from Mother Earth was so stimulating I realized I had found a niche where I could be comfortable for the rest of my life.

The second glide approach on finals made me feel as if I was sliding down a gigantic banister, which terminated just inside the boundary hedge. I mentally noted another Tigerschmidt rolling to a halt away ahead, to my right, and two other Moths on the takeoff roll, some distance to my left. This gave me a clear run into the centre portion of the grass 'drome. Just as I rounded out prior to touchdown, my attention was diverted by a large hare, which broke from the cover of a clump of long grass just a few yards ahead. The startled animal zigzagged wildly from side to side in a crazed effort to escape from my noisy machine, distracting my attention away from my usual reference point well ahead. The results were almost catastrophic! Instead of a smooth three-point landing, the long suffering trainer made two bone-shaking hops that would've done credit to a kangaroo. This deflated my ego somewhat as we came to a shuddering halt and turned 90° out of the prevailing wind, as dictated by airmanship rules.

I looked around and sure enough, just a few feet from the port wing tip, the brown and white hare lay in a crumpled heap of twitching, bloodied fur. Breaking a standing rule, which forbade leaving an aircraft engine running with no one at the controls, I quickly undid my safety harness and clambered from the cockpit. Picking up the injured animal by its hind legs, I dispatched it mercifully with a rapid chop to the neck, and deposited it in the baggage compartment beside the CFI's mushroom-stuffed helmet. Packing my machine and duly strapped in, as per regulations, I slowly taxied back to dispersal and shut down.

It transpired that I was the first student in our group to have gone solo, and no one had apparently noticed my episode with the unfortunate hare. Flight Lieut. Scott later thanked me for the unexpected, but welcome addition to his family meat ration, and discreetly avoided questioning me as to how this wild creature had found its way into the little baggage compartment.

My fellow students followed a time honoured after-solo ceremony by dunking me, fully clothed, into a deep static water tank filled with very cold water, close to the line of yellow painted Tiger Moths. But not even that chill could squelch my exhilaration. Even now so many years later, with over 50,000 flying hours logged, I can still smell that fresh air of the green English countryside intermixed with the smell of petrol and the unmistakable aroma of paint and dope-impregnated fabric which covered the old biplane wings.

I am glad that I took aviation as a career. My time just flew.

* *

5th from the right, second from last row

Just after my very first solo flight, training on the Tiger Moth biplane continued with short sorties around Peterborough, near the eastern coast of England. Since this area was within range of German fighter planes and night bombers, we were not allowed to venture far from our little, insignificant grass airfield. Hardly a suitable target for a bomber, but a tempting one for a fighter to strafe with an idea to kill trainees before they became qualified pilots. The powers that be finally decided it was a much better and safer idea to have the pilot training carried out somewhere else, in the extended British Empire. Quickly, South Africa and Canada were chosen, and hundreds of young airmen started on long voyages to these countries.

In early Autumn I found myself on a train headed for an aircrew holding unit on the outskirts of Manchester, Lancashire. It was a dismal place to be cooped-up in. Formerly an old hospital or asylum of some kind, it was surrounded by a very high and dark brick wall, topped with metal spikes. We were crowded into dormitories, with little space between the single iron beds, and with little to do, waiting for weeks with no information. Every few weeks a number of names would be called and these men, lads really, would be mustered away to crowded trains, destination unknown. Secrecy was priority number one for everyone. "Loose lips sink ships," was a watchword.

After several daunting and cheerless weeks of this confinement, I decided to have a little fun. Speaking to a number of my chums, I suggested a scheme which quickly met their approval. One afternoon, after a sparse lunch, I gathered thirty-six young trainees into a squad. By this time we were no longer AC2's, lowest rank in the Air Force, we were leading aircraftmen with a cloth two-bladed propeller above the elbow, on each arm of our tunics. Assembling right in front of Headquarters I called my squad to attention, had them fall in, in three ranks and right dress, to form three straight lines of twelve men. We then went through a drill session where the men would march forward, keeping in step, and then do an about turn on command. As the left foot hit the ground, I'd call, "Squad!" and then as the left foot came down again I would call, "About turn!" and they would reverse direction. We did this for about a half-hour, including practicing salutes and other maneuvers. Then I had the squad march around the driveway, directly toward the main gate. As we approached I could see the Corporal Service Policeman on duty there, staring toward us. I signaled with my hands for him to open the large iron gate, and did not slow our pace as we approached. The gate was swung open and we all marched smartly out without any query from the corporal, and carried on down the street, headed toward the underground station, out of sight around the corner.

Boarding a carriage bound for downtown Manchester, I made everyone promise to meet again two hours later, at the out-bound platform. We paired-off for a walk around the town. My friend Jim Yule, whose birthday was actually on Christmas Day, suggested we look for a N.A.A.F.I. (Navy, Army, Air Force Institute) Canteen, where we could get some hot tea and probably a sandwich or a bun. We were always hungry. We found one, packed with soldiers and some Air Force chaps who were attending a parachute jumping school somewhere nearby. Apparently they were dropping off a very high platform with a rigged 'chute, learning how to roll and fall as their feet hit the ground. Landing properly could avoid leg injuries etc…all very interesting!

After wandering through the streets rather aimlessly and popping into a department store, we headed back to the underground station. I was happy to observe that all thirty-six trainees were present. Marching back toward the main gate, I again signaled the guard to open wide and we marched in, feigning tiredness and a sort of, "Phew, glad to be back," air. No one ever queried what I had done. No questions. Nothing! A few days later, my group of about thirty men were called out of the morning parade and advised we would be boarding a train for Greenock, in Scotland, later that day, so we packed our kitbags and rallied when called upon.

The train for Scotland was jam-packed, and most of us had to stand for the entire journey, or squat upon our kitbags in long, dirty corridors with a toilet at each end. Occasional stops were made at towns which I had heard of but never seen, and we would sometimes be lucky enough to be handed a hot cup of tea on platforms where Salvation Army lady volunteers were doing their part for the war effort. The Sally Ann women (as we called them) were Angels of Mercy at times like this.

Upon arrival in Greenock we were quickly assembled, placed on a ferry, and motored out to a dingy-looking tramp steamer, anchored in the River Clyde. Kitbags were tossed into a huge net as we all clambered up a rope step ladder to the main deck. I got rust stains on my uniform, brushing against the steel plating of the hull of H.M.T. (His Majesty's Troopship) Bergensfiord. I thought it a dismal vessel, confirmed when I saw where we would be accommodated, a large square hold, amidship, and below the main deck. Two by four wooden struts lined the rusting bulkheads, and large hooks had been affixed for our hammocks. I wound up in a corner where mine was slung in the shape of a boomerang with too sharp a bend in the middle. I had to fall asleep bent almost double, in a "U" shape. It was dark and gloomy to say the least and whatever had been stored in that hold on previous voyages had definitely left strange odors behind. For two more days we lingered at anchor in the Clyde Estuary, before the anchor was hoisted and we finally started our long sea voyage.

Where were we headed for? No one knew. It could be South Africa, or Canada. We had all heard rumours about the training flights being cancelled in Britain, the fuel saved would be well used by Fighter and Bomber Command. We realised, as we stood on deck of the moving vessel, that we were now part of a large convoy. Lines of ships stretched for miles, and fast-moving destroyers of the Royal Navy constantly sped past as they searched for possible enemy submarines. Out of Scottish waters eventually, and wallowing past Northern Ireland, we headed out into a grey Atlantic Ocean. What followed was a monotonous fifteen days of drudgery, perpetually pitching and tossing in severe winter storms. Poor, almost inedible meals were often tossed up shortly afterwards by sea-sickness. I went up on deck as often as possible, with my service greatcoat wrapped tightly over my battle dress uniform, breathing deeply of the salty air, and peering through rain at whatever vessels I could discern through poor visibility. We seemed to be moving at the speed of our slowest ship, and it was heartening to observe sleek Navy ships doing their dutiful anti sub sweeps. Once or twice we did hear an occasional loud crumping noise, as some unfortunate ship was obviously torpedoed. This would be followed by numerous thumps, as depth charges were hurled overboard in attempts to nail the elusive submarine. It was with great relief one day, we saw a Canadian Air Force maritime bomber emerge from the clouds to circle the convoy and, the next day, early in January of 1942, we spotted land. After much maneuvering we sailed into Halifax Harbour, and eventually docked alongside dry land. Our sea journey was over. Where would we be sent next, I wondered.

Gathering our packed kitbags again, we all trooped down the gangway and headed for the largest railway train I had ever seen. It was parked dockside. The steam engine and cars seemed enormous compared to the British ones I had spent so much time in during the past year. All around us, generous ladies were walking around handing out hot cups of coffee or tea, biscuits, and sandwiches. We were quickly trooped into the spacious compartments still enjoying the goodies that were handed out. As darkness fell the train suddenly let out a loud "WHOOP" and we started off. Soon the outskirts of Halifax fell behind us and the train was making good speed through dark evergreen forests. The ground was covered deeply in snow, obviously freezing outside, but our compartment was warm and soon my companions were all stretched out, fast asleep. I however, sat in the corner seat staring out at the passing panorama of open fields and seemingly endless forests, interspersed with occasional level crossings where road and rail met. Destination this time turned out to be Moncton, in the province of New Brunswick.

It was still dark as we detrained and climbed aboard trucks painted in R.C.A.F. colours. Next came a short ride to a very large wire-enclosed campsite. This was Moncton holding unit, where all of those new arrivals in country would spend time until selected for onward passage to our new training facility. Our time spent here was very pleasant. The local population proved to be very generous in inviting us out to parties, where some of us learned to ice-skate, or ski. There were movies in the evening, or we could decide to try our luck by eating in some of the local restaurants. We had been surprised, shortly after our arrival, to hear about the Japanese December 7th attack on the American base of Pearl Harbour. America was stunned, but was now a full-fledged ally in our war. We were now brothers in uniform.

Eventually our happy time in Moncton came to an end. Some forty of us had our names called out, during the morning breakfast, and we were instructed to have our kit packed, ready to move out. We were all familiar with this routine, having seen many of our comrades paraded and dispatched in similar manner. Again, our unanswered question was, where to this time? The prospect of frozen prairie in Saskatchewan, or Manitoba, in a severe Canadian Winter, seemed ominous and indeed, unwelcome. However, we braced and were ready to move, come what may. A short truck ride down to the railway station, and once more we climbed into a large train compartment. Where were we bound, as our train shunted away from Moncton? I did not know, nor really care. All I knew was that, somewhere at the end of the journey my flying training would re-start, so I composed myself for whatever, wherever.

Our interest in the countryside was still high as we roared through numerous small communities. A stop was made in Montreal, where a military transport official ushered one group from the train before we started to move again. Meals were available, and drinks, both hot and cold, were plentiful as we gathered speed once more. I did notice that we were not heading due west at this point, but just considered that we were circumnavigating the city and its surroundings. The next town we stopped in however, was Toronto. We were not heading west, into the prairie lands, apparently. Next stop confirmed this fact. We were now in Windsor, Ontario, and about to cross the border into the United States of America.

In Detroit, Michigan, the group was ordered to disembark from the train, and then led to a large bus bearing the insignia of the United States Navy. All forty of us trooped onto this bus and were soon riding through the streets of Michigan's Motor City, en route to Grosse Ile, just outside the city limits. Again we entered a large wire-protected establishment, and were greeted by a huge Marine in dress uniform as we exited the bus.

"My name is Chief Maag," he announced, "I will be in charge of you lo for the time you are here."

His uniform sleeves were almost covered in stripes resembling sergeant's rank, but more numerous. This guy had apparently been everywhere, judging by the medal, ribbon, and other insignia on his chest.

"Follow me," he barked, as he strode off at quite a pace. We hurriedly shuffled after him, shouldering our kitbags and trying not to stumble. Our walk ended at a two-storied barrack block, where Maag informed us that we would be using the upper region of same. Upstairs we found a large room with forty beds equally dividing the space, twenty to each side. There was a metal locker next to each bed, where blankets, sheets, and pillows were neatly displayed. The wooden floor gleamed with polish, and not a speck of dust showed anywhere.

"This is how I expect to find this place every morning when I come to inspect it," the Chief muttered.

Quickly we all entered, and I managed to wrangle a bed close to the door with a window close by. This was followed by a sort of around the camp walk with a junior sailor who pointed out all the places we needed to know; such as the ground school, where we would attend lectures, and the mess hall, where we would eat our meals, sick quarters, where we would all undergo some hypodermic stabs, and the station cinema, where evening movies would be shown. All in all, a very tidy and well laid out camp.

"Watch out for Chief Maag," we were informed, "he's a real stickler for discipline, and he's worse ever since Pearl Harbour."

Within a couple of days we had all settled into a regular routine of "Rise and Shine, Breakfast, and Room Inspection," then off to ground school, marching in our own separate group at all times. We learned to walk upon large pieces of felt when in our dormitory. It kept the floor from being scratched by our heavy boots. In classrooms we started to learn more about the U. S. Navy, as well as the aviation subjects closer to our needs.

Everything was gearing up to full working pitch, as the Americans strove to overcome the former inertia of service life. We made friends in classes with young American flight students intent upon gaining the coveted Gold Wings. Next, we were all separated into smaller flights, and appointed a flight instructor. I think each instructor had four students.

My instructor was named David Taylor, an Ensign by rank, lower than a full lieutenant and only recently out of training himself, but keen and enthusiastic. Coming from somewhere in the Northeastern States, his voice was clear and concise. I liked him. Our aircraft was the U.S. Navy N3N Steerman, which was a lot bigger and heavier than the Tiger Moth I had soloed in, about twelve months previous. It was also fully aerobatic. We started off well, and I soon found myself at home in the cockpit. I hadn't forgotten my lessons in England and was soon progressing to more advanced maneuvers including aerobatics, and even some blind flying on instruments with my cockpit hooded.

After a few weeks at Grosse Ile, we found that we had just been processed through a grading course which weeded out the lesser-skilled students. Some of the British cadets were rejected and sent back to Canada for further assessment to new aircrew vacancies such as navigators, or radio operators, gunners etc. Those of us who had passed the grading were now to become students at Pensacola, the Navy's most famous training establishment. Late in February, thirty of us boarded another train in Detroit, headed south for Florida. We arrived at the small town of Pensacola, situated on the Gulf of Mexico, on a grand day of warm sunshine. Before us were golden beaches.

I thought I had died and gone to Heaven!

Once settled into the building allocated to us, we fell into the now-familiar routine of life in a Naval establishment. Our group was quickly outfitted with new khaki uniforms, short-sleeved, but retained our distinguished blue forage caps for identity reasons, as well as dress. Our food was plentiful, when in the mess hall, and well-dressed attendants hustled to fetch as much food as we could consume at each meal. We ate like Kings. After the meager rations served in wartime Britain this was ecstasy, and we all quickly gained weight. The classroom work proceeded at a faster pace, and our bus runs to Corrie Field became routine. Pensacola was so well-organized that a student could look at a chart and could see what stage he was expected to reach on any selected date. I took every challenge in my stride and sailed on, happy as a lark, day by day. We enjoyed odd days off and went into town often to enjoy banana splits, with three different kinds of ice cream in the dish, strawberry, chocolate, and vanilla. Somewhere along the line an unspoken decision was made for each student as to whether he would go to fly Fighters, or on to heavier types like the twin-engine flying boats called PBY Catalinas.

One day, after I had flown several other single-engine trainers, including the Harvard and the Kingfisher, I found myself down at the water's edge in the main base, staring at a flying boat called a P2YB. No name, just a floating hull with a stub wing sponson at the water line, and a huge, long wing with two engines, a float at each wingtip. I spent quite a few hours at the controls of this rather sluggish monster, flying over the blue waters of the Gulf of Mexico. We would take up several students for each long flight, and shared the time at the controls with the instructor. From the P2Y, we then progressed to the more modern Catalina PBY and more extended flights over water. Landing on the water in Pensacola Bay was a lot of fun but also hard work, especially taxiing when the waves were a bit rough.

Finally, in early September of 1942, I completed my training and prepared for the big passing out parade, due to fall on September tenth. That morning, with shoes polished and uniform shirt neatly pressed, I marched smartly with my R.A.F. cohorts toward the large parade ground. Brass bands were blaring out Sousa marches and single engine aircraft were flying over in formation as we all sprung to attention for the Commandant, Captain Case. I watched several young Americans step forward to have the Navy Golden Wings pinned upon their chest and be handed their official certificate, classifying them as Naval Aviators. When it came to the British contingent...my name was the only one called. I had come out top of my class, and stepped forward to receive my wings and certificate too. The Commandant was very gracious and wished me well, after congratulating me.

I walked on air afterwards.

When the parade was over our group was told to return to our ground school, where we would officially receive our R.A.F. cloth pilot wings. Assembled in a classroom and seated for a while, we all stood up as a Squadron Leader, who had flown in from Washington, entered the room.

"Please be seated," he said, "and I shall distribute your chevrons and wings by calling out your names in alphabetical order."

With that, he sat down at the teacher's desk and opened up his small attaché case. Extracting a pilot wing held to sergeant's stripes by an elastic band, he called out each name. Some he handed to people seated at the front, and some also walked a few steps to collect by hand. I was seated at the rear of the class and lo, when he called my name, he tossed the wings/chevrons at me. They fell short and landed on the floor. Yes, that was how I officially received my official Royal Air Force Pilot's Brevet. I picked them up off the floor of a dusty classroom. Without any words of congratulations, or further discussion, this wingless wonder from Washington closed his little case and walked out of the room. I was absolutely disgusted.

Later in our dormitory, I was busy sewing my wings and sergeant's chevrons onto my battle dress tunic, when I turned to find a row of guys saying they did not know how to sew, and could I do the honours for them? I wound up with blistered fingers, but the lads were all chuffed. Shortly after that we were all on our way north, bound for Canada. I dropped off en route to visit Molly and Irving Pitzak, the Jewish couple who had virtually adopted me while I was based at Grosse Ile, then back on the train again bound for Montreal. There I would commence a tour of duty with the Trans-Atlantic Ferry Command, based at Dorval. I learned how to fly twin engine aircraft like the venerated Dakota DC3, the B25 Fighter Bomber, the four engine, Liberator B24, and several other types.

Trips across the Atlantic were still considered quite risky. We often had to fly long periods solely on instruments, in extremely hazardous conditions, but it was all grist to the mill as far as I was concerned. Still, it was strange to be in Scotland one day, eating meager rations, only to be back in Montreal, eating chicken or huge steaks cooked to our liking, just a day later. I learned a lot during those five months in Ferry Command, and it has stood me in good stead ever since. My career was based on good training, good instructors, and a natural love of flying.

****

Late in the year of 1942, I was the captain of a Catalina P.B.Y. flying boat, which I had to ferry across the Atlantic from Newfoundland to Scotland. At the last minute before boarding, I was informed that we would have three female passengers. One turned out to be Martha Raye. The other two stars simply acknowledged me as a sergeant pilot, and immediately made their way to a couple of makeshift bunks at the rear of the kite. Martha asked me if she could sit up front. I did not have a co-pilot, so I agreed. She kept me laughing all night. We sang songs together, and she poured out my tea, or coffee, from flasks. That long trip was a real pleasure, and she kissed me when departing after we'd landed on the river Clyde, near Prestwick. I made a point of seeing all of her films, and watching her T.V. episodes. She was really a great lady.

DOWN IN THE DRINK

After five months at Ferry Command I was told to report to London from Scotland, where I would be informed of my next posting. This came as a complete surprise to me, travelling from Montreal with only the gear for my flight. All my personal effects and other items (my best blue dress uniform etc.) were still in my locker there. I was given a travel form to present at the railway and headed to Glasgow to see my mother. She was delighted to see me, and very grateful for the food parcel I carried. British rations were very short on solid foods, and the butcher shops would often run out of meat. My mother would make miracles with a bit of chicken or steak, feeding many from her pot. After enjoying a night at home, I headed to Glasgow Central Station the next day and boarded a very crowded train, bound for London.

After a rather frustrating day at London Headquarters I was told that, because the American training at Pensacola was weak in navigational experience, I was being posted to Number 3 School of General Reconnaissance at Squire's Gate near Blackpool, in Lancashire. Soon I was heading off in another train north once more, changing trains at Crewe. I was now in the company of two other Sergeant Pilots. We disembarked and trudged through the mid-evening dusk to find lodgings allocated by the Transit Officer who had met our train. The Grand Hotel was our destination (most hotels had been taken over by the Services) and finally it emerged from the blackout. The three of us headed directly for the dining room; we had been on crowded trains, for umpteen hours, with only a hastily-swallowed cup of N.A.A.F.I. tea en route, and were very hungry.

The dining room was empty, all tables bare. From the kitchen came the rattle of crockery and the clashing of pans being washed. I walked over to a hatch and knocked on it until it was opened by an irate W.A.A.F. Corporal, who informed me brusquely,

"Meal time is over," and down went the hatch, with a bang.

I knocked again on the shutter until it reopened, the same flushed face appearing.

"The three of us have been on trains since early this morning, travelling from London, and we've had nothing to eat all day," I implored.

Rather grudgingly, we were told to go and sit down. We would have to settle for some Spam and Chips, with a pot of tea to wash them down. When we finished I made sure to collect all dishes and utensils, returning them to the kitchen. Finding an empty room to settle into, I retired for a much-needed sleep.

The course at Squire's Gate consisted of straight cross-country navigational classes using charts and metal slide-rulers for measuring angles, ground-speed, wind direction and drift, etc. Flights were carried out using Avro Anson twin-engine kites. It felt strange to be up there doing navigational work instead of sitting at the controls, but it was all very instructive.

On our time off, and at weekends, we were able to visit the scenes at Blackpool Tower, and go dancing there in the evenings. By this time I had made friends with Corporal Bunnie Keelan, who joined us on many of our evenings out. I remember taking her to a cinema where Bob Hope, Bing Crosby, and Dorothy Lamour were appearing in, "The Road to Morocco," which I'd already seen in the States, but was happy to see again, with Bunnie beside me. We couldn't know it then, but she would stand by my side for the rest of her life.

When the course finished, I was sent to several different bases around England, flying some of the earlier twin-engine aircraft such as the Airspeed Oxford, and even the plodding old Avro Anson. Flying these types was a "piece of cake" as we would say. We were used to handling other operational machines in Ferry Command where, quite often, we were simply handed Pilot's Notes and sent off without any flight instruction. During all these months I kept contact with Corporal Keelan, writing letters, and meeting occasionally in London, where there were Service Clubs that had separate sleeping accommodations available for Service personnel.

Eventually I was posted to R.A.F. Silloth near Carlisle, on the border with Scotland. It was here I converted to flying the twin-engine bombers called Wellingtons, nicknamed Wimpeys, and selected the crew who would fly with me. This selection took place at a general meeting in a large empty hangar, where all the aircrew members gathered. We milled about, getting to know each other, and sort of weighing each other up gradually forming into six-man crews, Pilot, Co-Pilot, Navigator, and three Wireless Operator/Gunners. The Flight Sergeant allocated as my instructor was a bit dismayed to find that I had quite a few more flying hours in my log book than he had, but we got along quite well and I found his experience was useful, as we learned to drop practice bombs onto targets. This was followed by dropping dummy depth-charges on simulated submarines in the wide stretch of water that marks the border twixt England and Scotland. Low-level flying gave each machine-gunner ample time to strafe ground targets, with hundreds of 303 calibre ammunition.

Finally knitted together as a bomber crew, we were sent off to Talbenny in south Wales, where we picked up a brand new Wimpey to carry out service trials. When satisfied that all systems were airworthy, I was informed we would now be flying to Cairo, Egypt. Sergeant Bill Taylor, my navigator, and I put our heads together to work out our route, then submitted a flight plan for approval…Talbenny, direct across the Bay of Biscay, to Rabat Sale in Morocco for a night stop and refueling, then fly along the Mediterranean coastline toward Egypt, and Cairo. This all worked out quite well, with a short refueling stop near Benghazi, in Libya. We arrived eventually at a large base called Cairo West, about forty miles from Cairo.

It was right out in the desert. Heat was intense during the day and had us all perpetually in a sweat, yet sleeping in our tents at night required four blankets against the cold. Such a drastic change in temperature was startling, and quite hard to get used to. Much of our flying from this base involved daylight trips to harass enemy troops in the north Sahara desert, and night hops to bomb known fuel or ammunition dumps. We were given destinations along the Mediterranean such as the island of Crete; Greece, where enemy E-boats were operating, also received our attention.

We were now officially a part of the Middle Eastern Forces and expected to operate anywhere within that Command. Because of the shortage of aircraft in some places, my crew and I were often in demand. Given short notice I was one day informed,

"Head on down to Mombassa, in Kenya. They require a Wellington to carry out some anti-submarine flights to protect our ships heading for Aden, and the Suez Canal."

Next morning I went low-flying southward above the River Nile for hundreds of miles, gazing at all sorts of watercraft on its muddy surface. I flew over flat grasslands and saw all kinds of animal life. This was grand viewing indeed, a geography lesson with so many natural life scenes going by on the way to a stop in Nairobi, Kenya. Continuing on, we diverted well away from the Nile and headed due south for the east coast of Africa, and the busy port of Mombassa. The airfield was perched on a high plateau, well above the surrounding area, and its main runway terminated abruptly in a sheer cliff. Strong winds created odd air currents, requiring care when landing or taking off.

Armed with six huge depth charges, we took off next morning to patrol a given sea area, looking out for any signs of a Japanese submarine known to be somewhere in the Indian Ocean. This submarine apparently had a German U-boat commander aboard, training the Japanese commander in U-boat tactics. After several day-long sorties like this, we were moved a bit further north to be based at Mogadishu on the coast of Italian Somaliland, performing the same duties.

The airstrip here was just one fairly long runway parallel to the beach, and just above high-tide level. This was glorious to enjoy during our time off. We had cool sea breezes and even Italian ice cream, locally made in one of the tiny shops in the community. Again after only one week we headed for Aden, landing at Khormaksar, a permanent base for the Royal Air Force. All quarters were solidly-built brick structures. It was a very busy port, with shipping heading to or from the Suez Canal, north of the Red Sea.

The heat and high humidity in Aden was stifling and very difficult to adjust to. Body rashes from excess sweating and dehydration were common complaints. Getting airborne and up to cooler air was a boost to our morale. Crater City, a suburb of Aden, lies in an old volcanic cone above the city. We were often warned to stay clear of this area, because the Moslem Arabs who lived there were actively opposed to strangers. Our duties when attached to 621 Squadron at Khormaksar were to carry out anti-submarine patrols, and to keep a weather eye upon any shipping in the area.

On January 4th 1944, we took off on such a patrol, our aircraft outfitted with special long-range fuel tanks. These would keep us in the air for at least ten hours of flight. Heavily laden with this fuel and our six depth-charges, we staggered off the runway and headed south-east over the Indian Ocean. Several hours later and well out to sea, I requested that my navigator switch over to the reserve fuel tank.

"Make sure that you operate the correct balance cocks," I reminded him, and carried on searching the sea and sky around me.

"Okay Skipper," he replied, "Tanks switched," as he returned to his desk.

A brief minute after that, the port engine suddenly coughed and died. I corrected for the swing and checked my instrument panel. All seemed in the green. As I was about to have the navigator do a check on the selectors, the starboard engine also failed and we commenced to lose height. We were flying at 2,000 feet and dropping fast. I told "Taffy" Thomas, my wireless operator,

"Send an S.O.S. and then get to your ditching position. Crew, standby, we are about to ditch."

I was also turning into a strongish wind, busily trying to restart an engine at the same time, but eventually had to flare off my speed as I headed along a crested wave. I selected 15 degrees of flap, to lower my touchdown speed, and held the aircraft just above the wave in a tail-down attitude. The rear turret touched the water and we perceptively slowed as I pulled the elevator control back to its full extent. More slowing down was felt, and then the nose dropped.

It was like running full-tilt into a cement wall. Bang!

I was thrown forward against my tightened harness, felt myself being tossed against something hard, and then fell flat on my back, wondering where I was. Lying there, I was looking up at a light green rectangle perceptively deepening in colour. With a start I realised that I was laying on the bomb-aimer's panel (which was also our door in and out of the aircraft) looking up at the jettisoned hatch above the pilot's seat. The aircraft was obviously sinking fast. I quickly released my parachute harness, struggled out of it and climbed up, using my empty seat as handhold. My head popped above the waves just as the four guns in the rear turret sank out of sight.

The round orange-coloured Lindholme dinghy had operated out of the starboard engine nacelle, and I could see some members of my crew were already aboard. I swam to it and clambered aboard, counting heads as I did so. Four heads, with myself making five. Who was missing? I was told that my wireless operator had remained at his post, sending out an S.O.S. right until we splashed down. He should have been at his ditching place. As one, we all held hands and uttered a prayer for our lost comrade, then started to take stock of what we had rescued from the aircraft, giving thanks at the same time for the number of times we had practiced the dinghy drills, even at night in pitch darkness.

It was full daylight still, and the heat of the sun was fierce. Our saturated clothing soon dried, and we started to get sunburned. Night fell, and in the darkness we could often see phosphorescence in the waves surrounding us. The sea got rougher and the dinghy was tilting quite a lot, even though we spread out to even the loading. I was sitting up on the rounded roll of the dinghy wall, when suddenly the sea was in my face. The dinghy had pitched over and we were all in the water, swimming while trying to hold onto the pitching life raft. Using the wind to help us, we managed to turn the dinghy over and get it right side up. We climbed aboard to find that all of the loose equipment and rations had been washed overboard. Nothing was left. This was serious.

During a careful inspection of the raft for leaks or tears my tail-gunner, Bob Scott, noticed a cord dangling below the rim on his side of our floating refuge. He took hold and pulled it in. To our great relief it turned out to be attached to a container housing several bottles of water, some Horlicks tablets, and a Verey Pistol, plus a dozen cartridges for emergency signaling. What a prize! I warned my crew that we would have to carefully ration our drinking water and hope for some rain, maybe. As the day warmed up, we again felt the assault of the sun. Our skins started to blister. For a while several shark fins were visible as they circled our frail support and, on one occasion, a shark swept below the dinghy, its rough hide making a rasping scrape as it passed. Eventually though, the heat forced us to immerse our bodies, one at a time, into the sea to obtain fractional easing. When one was overboard, all other occupants maintained a close watch, and if a dorsal fin veered closer we would start thumping on the taut floor of our raft. That seemed to scare them off.

Late that afternoon we heard aircraft engines, far off, and voiced our thoughts. Perhaps a search was being made for us. That brightened our outlook. Eventually we actually saw a Wimpey, flying at about 2,000 feet. I loaded the Verey Pistol and fired a couple of cartridges. They soared high above us, but the searching kite just flew blindly onward and disappeared. Our second night fell and we were all rather despondent. Crouching together back-to-back in the centre of the dinghy, we managed to keep it flat during the pitching and rolling of the night. The canister, with its precious contents, was hugged tightly to my chest at all times.

The next day, our third exposure felt even longer and we all spent longer times fully immersing our bodies in the sea for relief. The sharks did not seem to be such a menace anymore. Twice more we heard aircraft flying nearby and more Verey cartridges were wasted, without result. Again a Wellington bomber appeared, this time much closer. He must surely see these I prayed, as I fired more of the precious signaling projectiles, but no such luck. I wondered if they were all blind, or asleep. Now we were down to just a couple of cartridges. We had to be careful when, and how, they should be fired.

The third night fell to give us surcease from the blazing sun, and it was my time for watch. The sea was relatively calm and I was seated atop the rounded outer air bag when I saw a bright flare of light, lasting only a couple of seconds. I immediately pointed the flare gun upward and pulled the trigger. A red flare soared high. All at once my crew was awake and asking me questions. I told them I had seen a light, and pointed in the direction. All was dark. We had a short discussion about false alarms, but I was convinced that I should fire our last cartridge. Exercising my position as Captain I reloaded the pistol, and with some misgiving, pulled the trigger. Our last cartridge flew high, and then faded.

Meanwhile, aboard the H.M.S. Lulworth, a young sailor who only meant to sneak a forbidden smoke behind some depth-chargers near the stern of his ship, was speaking to an officer.

"Sir, I just saw what looked like an emergency flare in the sky!"

"Where?" demanded the officer. As the sailor indicated the direction, a second flare soared high.

"Look sir, there's another one!"

Quickly assessing the bearing, they advised the bridge and the ship changed course, heading our way. Within a few minutes the warship drew alongside our crowded dinghy, and we were climbing a ladder to be welcomed aboard. What fantastic luck! We owed our lives to a sailor stealing a quiet smoke, supposedly forbidden when on deck at night. Our savior Royal Navy vessel was escorting a tugboat, which was busily towing a Liberty ship that had been torpedoed, but didn't sink. They were headed for Aden, the nearest port. Strict radio silence was being maintained at the time, in case signals alerted the enemy. No position reports could be made. Five days later the port of Aden was visible and messages were sent by signal lamp, informing authorities that the ship had rescued air crew on board. This was the first knowledge for the air base that we were alive. Messages had been sent to our families that we were missing in action and feared dead.

Once secured alongside the dock, we all said goodbye to the Captain and crew of the Lulworth, and thanked them again for our rescue. We met the sailor who had seen my first flare, and kept secret it was his act of lighting a cigarette that prompted our rescue. A very ancient, decrepit ambulance was standing by on the quay. We were all ordered to board and lie down in stretchers, though by this time we were hale and hearty, fully recovered from our three-day ordeal at sea. The Navy had looked after us well, even dressing us all in white shirts and shorts. The military hospital sat atop a high hill overlooking the port and was reached by a very winding road which twisted and turned tortuously as it climbed. Rounding one such bend where the road tilted right to left, the ancient and large framework of the ambulance contorted, allowing the back door to flop open. The stretcher I was lying on slid backward, banging me down head first onto the tarmac. My crew yelled out for the driver to stop, which he eventually did. He came running back to me, uttering in a broad Scottish twang,

"You'll no report me sir, will ye?"

Reaching the hospital, we were greeted at the entrance by a busty Matron who ushered us all into a ward and commanded us to bed. Our Wing Commander Flying, and the Squadron Leader both arrived a short while later, and asked to talk to us. The bossy Matron ordered them off the premises, saying they could speak to me the next day, after I had rested. Once back at the base we told the whole story of our flight, and how both engines had cut out. Apparently another Wellington had ditched a day after we had, and under similar circumstances. Suspicion had fallen on one Arab who was operating the refueling truck at the base. Apparently he was putting something other than aviation fuel in the long range belly tanks.

A short while after we returned I was flown from Aden to Benghazi, via Cairo. I joined 38 Squadron and took over a new crew doing night operations over Italy, where most of the mid-east action was now centered. A month later I was posted to Ein Shemer in Palestine, as an instructor. Based near Haifa, I was able to travel quite widely in the Holy Land. On weekends I visited Jerusalem, rode on horseback near Bethlehem, and spent time in contemplation, walking along the shore of the Sea of Galilee. After eighteen months in the Middle East I was flown back to Britain, eventually going through the conversion course to fly Lancasters in Bomber Command.

ROYAL AIRFORCE BASE

The weather had been terrible for three days with torrential rain, low cloud ceiling, poor visibility and of course, no flying. Although grateful for this respite from night operations over enemy-occupied Europe, most of the air crews were cheesed-off and beginning to feel restive.

"Our flight commander had us practising ditching drills in the Lancaster mock up, over in Hangar One," moaned one Sergeant, "He had us doing the bleeding drill over and over again even in pitch darkness, until we had the time down to just thirty seconds for everyone to get out, complete with all our emergency gear."

"Quite right too," one of his listeners stated, "could perhaps save your miserable life one of these nights, if you ever have to splash down in the North Sea. These Lancs don't float too long, especially if they've been shot full of holes."

In fact most of the aircrews had been made to use this down time to brush up on all emergency drills. All the newest squadron aircraft had been taxied out to the test range, where the air gunners synchronised all machine guns to the approved convergence range. This was supposed to shred any enemy fighter caught in their deadly hail of bullets. Wireless operators and navigators cheerfully attended classroom lectures to brush up on the latest operational radar, and radio equipment. Such knowledge could mean the difference between life and death in the aerial war. The less experienced pilots practised blind instrument flying in the station Link Trainer, knowing that soon they'd be called upon to fly out of sight of the ground, without a natural horizon.

Of course all of the ground crew personnel, especially those who actually serviced the aircraft, had received no respite during these bad weather days. Everyone was occupied with repairing damaged machinery, and performing minor miracles to restore all Lancasters to flying standards. Sometimes these tasks were carried out in the relative warmth and comfort of a closed door hangar, but for most of the riggers, fitters and mechanics, it meant slaving away under atrociously uncomfortable conditions out in the open air, with rain and occasional snow flurries to further harass them. These tradesmen and women were the unsung heroes and heroines who formed the backbone of the Royal Air Force. Without their stoic endurance and devotion to duty, the bombing offensive would have quickly ground to a halt.

This Thursday morning in January dawned brighter, with cloud cover beginning to break. It lifted everyone's morale somewhat, and it wasn't long before a buzz of expectation shot through the base.

"What's the betting that we'll be on Ops tonight?" my navigator Bill Taylor ventured.

I didn't take him up on what was bound to be a sure thing. Shortly after breakfast, the station tannoy system blared out its message.

"All squadron crews report to their respective flights."

When I checked in at 103 Squadron with my Flight Commander, Stan Matthews (nick-named after the celebrated football star) I was informed that my Lancaster, "F" for Freddie, had been repaired and I'd be flying it that night. The night landing from our previous operation had been rather scary. Unknown to us, our left tyre had been holed by anti-aircraft fire. The port undercarriage leg collapsed near the end of the landing roll because the wheel jammed, causing us to veer off the hard runway surface. The port wingtip and one propeller were badly damaged.

"You'd better nip out before lunch and do a full air test on it, and for God's sake try not to bend anything else. We're still waiting for a replacement for "C" for Charlie, which pranged last week," said the Flight Commander.

Out at the concrete dispersal point, I completed the walk-around check of my favourite kite. Everything seemed to be tickety-boo, as good as new. After carefully checking that previous snag entries had all been officially cleared, I signed the Form 700 and boarded with the rest of my crew. The air test went well, all systems checked out properly and we landed without incident, leaving the various ground crew tradesmen ample time to prepare for the forthcoming night operation. Of course by this time almost everyone had been informed that the station was closed. No one was permitted to leave the camp, and no outside telephone calls could be made.

At times we aircrew thought that these local security measures were a bit foolish. The local population (who were by now familiar with our traffic patterns) could easily detect when an operation was about to be launched. Absence of familiar faces in the local pub in the early evening and frantic movement all around the airfield as bomb trolleys, fuel bowsers, and other assorted vehicles clattered back and forth, all let them know. The increase in aerial activity during the day when aircraft systems were tested were all a dead give-away too.

Any astute aircrew member could of course find out how much fuel had been pumped into the waiting Lancasters by asking the Bowser drivers, and similar queries to the Armourers would elicit what sort of bomb loads had been trolleyed out. These figures could give a navigator or pilot a rough estimation of the radius of action each kite could fly. I had made these fuel / bomb enquiries shortly after completing the recent air test and realised that it would be a long night's work but of course, kept those thoughts to myself. As long as the details of the actual target destinations were kept secret, we were relatively safe. The evening briefing would eventually reveal all.

The atmosphere in the Sergeants Mess was a lot more congenial than when I had first started flying as a Sergeant Pilot. Pre-war messes had tended to be rather sedate, with ground crew Sergeants, Flight Sgt's, and the occasional Warrant Officer residing in a clubby atmosphere, suitable for servicemen of long standing. Having joined as Boy Entrants, or chosen to volunteer before 1939 for a career in the Royal Air Force, they had all worked their way up through the ranks, taking years to reach their present status. The promulgation that all aircrew students would be promoted to the rank of Sergeant or commissioned upon completion of their training syllabus, and the awarding of the appropriate aircrew brevet (be it a full pilot's wing, or the half wing of other trades) did cause some resentment in the regulars.

"Bloody jumped up Sprogs," they'd mutter, "got three stripes in a matter of weeks, not years."

Of course with the average age of the new aircrew as low as 18 or 19, this effect was further aggravation to the older Mess members. Resentment at times was quite marked, especially when the high-spirited youngsters started playing pranks and games guaranteed to rile them. When one young Sergeant told a Warrant Officer that Prime Minister Winston Churchill had called aircrew members "The cream of Britain's youth," the Oldster replied, "Yes, whipped up into Clots!"

Eventually though, by observing aircrew losses and beginning to understand the risks involved when flying night raids, everyone started to acknowledge each other's full contribution to the war effort. The Sergeants Mess became a much happier place for relaxation and fun when off duty. We were no longer greeted with expressions like, "Get some in," meaning service time.

After lunch I advised the crew to follow my example and snatch a few hours rest. Naturally this wasn't exactly easy. One couldn't help wondering what sort of flight lay ahead. Would it be an easy target, a "piece of cake" as we'd say? Or would we be flying deep into the very heart of Germany? I had a good hunch already. Veteran crew member or raw rookie, we all experienced the same doubts. Would we make it back safely? Who would die tonight? What were our chances of completing a full tour of thirty op's? Holding the coveted rank of Warrant Officer, I was already halfway through my second tour with an awarded nickname of "Lucky" and regarded as a veteran because my first tour had been on twin engine Wellingtons, nick-named Wimpeys. Despite the nagging uncertainty and nervous fluttering in my stomach, I managed to drop off for a dreamless couple of hours or so, before being awakened by the Station tannoy's insistent blaring.

"All crews report to the Operations Room for briefing."

I quickly joined the swarm of young men noisily streaming toward the long Nissen hut, forming naturally into crew teams as they gathered. My crew consisted of Sergeant Bill Taylor as Navigator, hailing from Liverpool. Sergeant Ken Fuller, Flight Engineer, was a quiet spoken Canadian from Vancouver. Flight Sergeant Dusty Miller, who came from a small village just outside Folkestone, Kent was responsible for Bomb-Aiming, and the two machine guns in the nose of the aircraft. My two Air-Gunners, both Sergeants, were Jiggsy Giles from London, manning the mid upper turret with two Browning machine guns, and John McVitie from Scotland, whose domain was the four-gunned rear turret. Jock had a thick Scottish burr to his voice, with a Glaswegian accent broad enough to walk on. Last but by no means least, there was my Wireless Operator, Flight Sergeant Paddy Baylor from Belfast. As a WOP/AG he could double-up as a gunner in either turret, should the need arise.

In the briefing room rows and rows of seats and trestle tables faced a raised dais, where two wooden easels supported a large cloth-covered map. The chalk boards on the rear wall displayed a list of the aircraft call signs, plus the names and rank of the pilots who would command them. The walls were festooned with black and white silhouette drawings and photographs of enemy aircraft. Several wooden aircraft models dangled from the curved ceiling as recognition aids. We were a motley bunch of young men, from many different parts of Great Britain and the Commonwealth—Canada, Australia, New Zealand and South Africa were pretty well represented. Different backgrounds in education, social standing, and religious beliefs maybe, but all with one common aim—hammer the enemy and get home safely in one piece as quickly and as often as possible!

A sharp call of "Attention!" brought us all to our feet as the Station Commander marched in. He was followed by the Wing Commander Flying, the Intelligence Officer, and the Meteorologist who was, surprisingly enough, a W.A.A.F. Officer. "Be seated gentlemen," Group Captain said, and there was a noisy shuffling of feet and scraping of chairs, taking some time to diminish into expectant silence. Everyone's eyes were now affixed to the shrouded rectangle on the easels as the Group Captain strode forward to remove white sheet from the target map.

There was a loud buzz of excitement and some indrawn hissing of breath as the map of Britain and Europe was revealed. As I had suspected, there were long red streamers leading from numerous airfields in a series of dog legs, converging and terminating deep inside Germany.

"Gentlemen, the target for tonight is Berlin. There will be an Armada of 1,000 bombers in the air, and we are going to clobber the enemy where it will hurt most." He then nodded to the Wing Co Flying to take over.

Using a long pointer, the Wing Commander traced three red lines on the large map. The first and longest leg took a route across the North Sea to a rendezvous point near the Danish coast. The second red line ran inland on an Easterly heading to an easily-identifiable ground location. Third and final line pointed directly to Berlin in a South-Easterly direction.

By following this route, it was hoped to split the enemy air forces and delay their confirmation as to our actual target area. Night fighter squadrons usually were not launched against our bomber streams until the German radar sites plotted the tracks of each separate R.A.F. formation. Other squadrons of bombers would make subtle feints further south to add to the confusion of the enemy defences and split the strength of the opposition. "Well that's the theory anyway," muttered one disgruntled air gunner who had faced the night fighters a number of times. Of course we all tended to see this routing in a different light. The fancy dog-legging meant that we would be in the air longer and be over enemy territory for much of the time, especially on the slightly more direct route homeward bound. Consequently, we would be exposed to night fighters and concentrated anti-aircraft for quite a long time.

Newer crews were busy taking copious notes of these details. The briefing continued with the Met Lady explaining weather expected en route and for the return flight, which everyone present devoutly hoped they'd be making. This was followed by the Intelligence Gen, with updates on known areas of heavy flak concentrations, and expected strength of night fighter aircraft.

"The Path Finders will be ahead of you and will mark the targets with flares. Take careful note of these colours for targets, because the Germans have been known to drop flares of their own to mislead us."

Despite propaganda to the contrary, our specific targets were all legitimate military objectives—munitions factories, airfields, manufacturing plants, and intricate railway junctions—all easily identifiable near the lakes, or other water surfaces that could not be camouflaged successfully from aerial reconnaissance if the sky was clear, as forecast. It was a sad fact that the Germans quite often constructed some of these important military targets well inside civilian conclaves in an effort to disguise them, and then claimed that the Royal Air Force was insensitive when the civilian population suffered during bombing raids. Weren't they first to drop high explosives on civilian targets from the air?

Details followed as to altitudes allotted to each group of bombers, time of take-off, time of arrival over each rendezvous, and expected time over targets, plus a reminder to set aircraft altimeters to a standard setting once above 10,000 feet. This was supposed to ensure vertical, as well as horizontal separation between the converging four engine attackers. A period of questions and answers was then followed by the ritual which always convinced me that indeed, we were going to be committed to flying an actual night operation. The C.O. in a loud clear voice announced,

"Gentlemen, synchronise your watches." As he gave a ten second countdown to a stated minute, everyone present pushed in stem of their service watches. After a hearty, "Good Luck Men," from the senior officers, we all rose as they trooped out of the room.

Having eaten a good pre-flight meal attended by all crew members, we assembled in the squadron changing rooms to struggle into our various layers of clothing. This was usually accompanied by ribald comments, corny jokes and a sort of affected hilarity which couldn't quite disguise the electrical nervous tension, hovering almost visibly in that crowded locker room. There were plenty of reminders to empty our pockets thoroughly, and not carry anything which could prove informative to the enemy, should anyone be unlucky enough to be shot down and taken prisoner. Except for assorted good luck charms and personal amulets or photographs, this rule was very strictly enforced. My affectation was a white silk scarf around my neck. This had been given to me by a girlfriend when training in America. I wore it to prevent my neck from being chafed against the collar of my uniform when swivelling my head from side to side, searching the sky.

All too soon it seemed we were gathered outside, laden down with our gear and awaiting the drivers who would run us out to the waiting Lancasters in the far dispersal areas. Our driver turned out to be a rather busty W.A.A.F. Corporal which prompted Jock McVitie our rear gunner, to comment.

"I wonder if they gi' these lassies a different shaped 1250 tae put in their tunic pockets?" 1250 was the number given to our official identity cards, usually carried in the left breast tunic pocket. Another remark was, "She's built like a Beaufighter," referring to the famous twin engine night fighter of the time. It had one massive radial engine projecting forward from the wing on each side of the fuselage.

These service women were pretty well used to this sort of clumsy badinage, and fully understood the personal doubts and tensions we were all undergoing. The so called "Pre-Flight Jitters." Sometimes they gave as good as they got, with snappy back answers that could deflate anyone who made observations much too personal. Quite often these ladies and other ground personnel would stand for cold lonely hours, waving us off on our trips and if personally involved with a crew member, would also wait in the early hours for that aircraft's return.

During my pre-flight walk around, I carefully examined the contents of the opened 33 foot-long bomb compartment. A mixed load: one 4,000 pound bomb nicknamed a Cookie, and 12 x 500 pounders. These plus 2.154 gallons of fuel in the wing tanks, and fourteen thousand rounds of ammunition for the machine guns brought the Lanc's weight up close to maximum. It was customary now, with time to spare, to have everyone enter the aircraft and recheck all systems. I ran up all four engines of Freddie and carried out magneto checks, once all the systems temperatures and pressures were "In the Green," before shutting down. Every member of the crew then piled out onto the concrete for the final nail biting minutes. Those who smoked moved away a short distance downwind from the heavily laden and highly inflammable kites, while the rest engaged in sporadic conversation with the ground crew, patiently waiting for their charges to trundle off into the darkness. I usually took this opportunity to let these ground personnel know how much we appreciated their efforts to keep our aircraft in tip-top shape, often while working in very adverse conditions. We couldn't do our job without them.

Next came a ceremony which often puzzled onlookers. Gathered by the tail of many Lancs, the assembled crew members each carefully unzipped their flying suits and fumbled inside. A few seconds later seven streams of urine fountained and splashed over the tail wheel. This was not just an affectation or a sort of boyhood prank, but a necessary before-flight relieving of ones bladder, to hopefully offset the need for doing so whilst at high altitude and cooped inside the freezing fuselage of the bomber. I sometimes wondered if this regular christening ever had any deleterious effect on the long suffering tail wheel. One evening, just as we were about to perform this ceremony, one of the corporals standing by the trolley acc. yelled out, "Preeee-sent Arms!" in a Sgt. Major's stentorian voice. It really had us in fits!

Consider that each man was now wearing a heavy canvas outer flying suit over bulky kapok-lined inner suit, on top of service blue battle dress trousers and blouse. Under these three layers were three more: a shirt under a heavy woollen sweater, and thermal long john underwear next to the skin. The Canadian air crews called these Stanfields, after the name of the manufacturer. On top of all these items each man had donned a Mae West lifejacket and a parachute harness, plus flying helmet, goggles, and gloves. Our feet were also encased in thick wool socks inside bulky fleece lined boots, for warmth at altitude. These clumsy boots had a nasty habit of falling off if one had to bail out. Try fumbling to the core of one's being when encased in all of those layers, and a sub-zero wind whistling in to invade your most intimate and private parts!

These were the reasons why, while standing upon the comparative safety and comfort of terra firma (the more firma—the less terror!) those long suffering tail wheels were so ceremoniously deluged to become an integral part of each night flight. It was a special moment of togetherness each crewman cherished. It said, we are a team, and we are ready to share the dangers of the night—let's get this over with!

Suddenly the darkness of the evening was broken by the firing of a flare from the Air Traffic Control balcony. That was our signal to start engines. The operation was now definitely on. Up until this moment quite a few of us had secretly hoped that the operation would be scrubbed. A couple of minutes later, I signalled to the ground crew "Chocks away" and, with a roar of four Merlins, F for Freddie lurched forward onto the perimeter track to join a procession heading for the runway threshold. Two miles away in the village, the local population would now be aware that their adopted lads were on the way. They could hear wheel brakes squealing, and then the roar, as each heavily laden leviathan soared overhead, to be swallowed up in the darkness. Many of them, I am sure, uttered a prayer for our safe return, as we ourselves did.

Taxiing past one dispersal, Jock in the rear turret mentioned that one of the Lancasters, Q for Queenie, was still sitting there with no engines running. The crew were standing around it like spare guests at a wedding. I told him Ned Sparks was the skipper of that kite. They had probably developed some snag and would soon catch up with the rest of us. Finally arriving at the holding point just short of the runway threshold and awaiting clearance from the control, I checked communications again with each crew member and ran through the pre take-off drills.

"There's the green lamp, Skipper," my upper gunner reported.

"Roger. Pilot to crew, standby for take-off."

Once in position on the runway I increased the power slowly, holding the brake pedals depressed until the entire machine trembled and shook. Good brake pressure was assurance that our landing run could be slowed down safely. I throttled back to idling power, then pushed all four throttles forward. I ordered the Flight Engineer, seated beside me, to take over the four throttles and give me take-off boost, monitoring all engine temps and pressures. He replied,

"I have the throttles Skip, and will go to full power on the roll."

I placed both hands on the wheel as F for Freddie continued its long roll into the darkness. As speed increased, the tail lifted and the control wheel started to vibrate, emphasising that my kite was coming alive. She'll let me know when she's ready to fly, I said to myself, as the end of the runway drew closer.

Just above 90 miles per hour, with a gentle back pressure on the control yoke, the main wheels left the tarmac. The trembling and rumbling noises decreased. We were airborne! Trees and houses flashed past beneath our wings. I applied light foot pressure on the brakes to stop the wheels revolving.

"Wheels up please, and give me climbing power Ken."

The mission was on its way! Strangely enough, any misgivings I may have had previously were now relegated to the back of my mind. I felt no fear and could simply concentrate on handling my machine. The other crew members felt the same way, I'm sure.

Securely strapped in with shoulder and thigh restraints, I concentrated upon the flight instruments, turned onto the desired compass heading provided by the navigator, and ensured that both climbing speed and gain in altitude were correct. Ken Fuller my Flight Engineer, rechecked all engine temperatures and pressures. It took several moments before he could get the four propellers properly synchronised. The angry offbeat thrum of Merlin engines eventually settled down to the sweet smooth hum that gave this sturdy Lancaster its own peculiar signature tune, readily recognisable to ground listeners, friend and foe alike. The intercom buzzed,

"Bomb Aimer to pilot, we are now crossing the coast outbound."

"Roger Dusty." Still below cloud, I advised the bomb aimer and both turret gunners to test their guns.

"But for God's sake, don't shoot until you are certain the immediate area is quite clear. Everybody keep your eyes peeled. Use the clock system and let me know if you see any other kites."

For a few brief moments all eight guns chattered, and the distinctive smell of cordite permeated the fuselage.

"Guns O.K. Skipper," all three reported in unison, just before we entered cloud for the long climb.

When the altimeter showed ten thousand feet I reminded everyone,

"Ten-thousand feet. Check oxygen supply is on."

One by one they all reported in. Having acknowledged each, I reset my altimeter to the standard setting advised during the briefing. The theory was, if every pilot accurately maintained his given altitude, there should be less chance of a mid-air collision. This of course did not rule out having two bombers converging upon each other due to slight differences in airspeed and compass variations. In the dark confines of the cloud cover it was impossible to see the rest of the formation, but occasionally I felt a tendency for a wing to dip as we traversed across someone else's slipstream or wing tip vortices. This was the spooky part, knowing that an entire squadron of heavy bombers were all climbing through the same air space and heading for the same rendezvous, though not another one was visible so far.

Most often this climb to altitude would have been confined to clearer skies inland before heading out toward the North Sea, but the length of this night's flight precluded adding extra mileage to an already formidable total. The weather front which had grounded us for the previous three days had now moved, causing the clouds to break into layers and thin out somewhat. Eventually at 14,000 feet, F for Freddie popped into a clear starlit sky, leaving us feeling very exposed. For the first time since shortly after take-off, I was able to spot other aircraft dotted around us, all of them climbing to operational altitude.

Stubby exhaust pipes glowing red hot on each side of the powerful Merlin engines were a dead giveaway and even the slight glow from a crescent moon, although an aid to attitude control by giving us a horizon to look at, was resented. It could give the enemy night fighters an advantage we'd rather they didn't have. In the station cinema not too long before, they had shown a British film called "Dangerous Moonlight," starring Anton Walbrook, and I remember thinking how exposed the Wellington night bomber had seemed in it. I felt exactly like that now. In later engine models these exhaust stubs were all shrouded. Cloud cover would have been much more preferable.

Throughout each leg of the flight Paddy, our wireless operator, maintained a constant listening watch on pre-selected radio frequencies. This was in case headquarters issued a recall signal (It had been known to happen) but no such message was received. We were still "Go" and now a unit in the miles-long stream of bombers. Almost two hours after take-off we were flying level at 22,000 feet, with the outside air temperature reading a chilly 25 degrees below zero, clocking along at an air-speed of 210 m.p.h. The chill factor caused by numerous draughts was horrendous.

"Navigator to pilot, we are approaching the first turning point. In two minutes we change heading to 095 degrees compass."

"Roger Bill. Acknowledge 2 minutes and then steer 095 degrees. Captain to all crew. We are approaching the enemy coast. Everyone stay wide awake and keep your eyes moving. Watch out for night fighters."

One by one the keyed-up crew members responded with an

"O.K. Skip, will do."

I could feel Freddie's fuselage wobble and yaw slightly as the tail gunner swept his turret from side to side in a standard sky searching pattern. At this point I remembered an occasion when one of my wireless op /air gunners had been asked, "What's it like away up there?" He'd instantly replied, "Cold outside—bloody freezing inside!" To prevent frostbite we had all been issued a pair of white silk gloves, to be worn next to the skin. Over these we also wore a pair of woollen gloves. Each of these had separate fingers to facilitate touching and operating individual switches or buttons. At the controls I also wore a pair of thick brown leather gauntlets with a separate thumb to ensure that I could grip the control wheel firmly. An outer flying suit of later design had been produced, with electrical wiring throughout, which was supposed to draw current from the aircraft system and provide heat for each crewman's body, but they seldom worked satisfactorily. The cosiest working position in some Lancs was the Wireless Operator's, because of a duct which diverted warm air from the Merlin engines. On occasion he had been known to complain of being uncomfortably warm, while the rest of us had frozen extremities.

In the meantime, as always, I was concentrating upon flying accurate headings, heights, and airspeeds. Just few minutes later we crossed the enemy coast and the anti-aircraft fire started to pock mark the skies around us. Searchlight beams swept to and fro, seeking out sky-borne invaders. Multi-coloured ribbons of incandescent light wormed their way skywards, as high explosive shells headed toward our elevation. For a while they seemed to be coming straight toward us, each one looking to be a direct hit and then at the last minute, because of our speed, they would arc away to explode in a black cloud of smoke above, below, or uncomfortably close to us. It was a sort of hypnotic display of deadly fireworks which could easily destroy us. Occasionally the aircraft would buck as it passed through an area of near explosion, and then it would sail serenely onward. The sky always seemed to be full of stinking black smoke. I've often heard "Line-Shooters" describe it as being thick enough to walk on!

Ahead and to starboard there was a sudden bright flash. One of our aircraft had been hit, probably in a wing fuel tank. With its port wing ablaze it started to spiral earthwards. "Get out. Get out. Hit the silk!" I was mentally willing the endangered crew, when there was a tremendous fireball as its bomb load exploded. It literally blew the stricken Lanc to smithereens. Moments later the concussive shock wave hit us and F for Freddie rolled almost 90 degrees to the left. As I regained control to level flight I called,

"Did anyone see parachutes? Did anyone get out?" The answer was a depressing,

"No, Skipper."

A couple of minutes later, from the mid upper turret, Jiggsy (who had the best all-around view) suddenly screamed,

"Enemy fighter at 5 o'clock level. Corkscrew— NOW!"

Going through our well-rehearsed drill, I immediately yanked back on the control column and banked steeply to starboard, turning in toward the fighter. She shuddered, but reacted well in zooming skyward. Moments later I was reversing the procedure and diving to port, thus increasing the fighter pilot's deflection for shooting. This was our best defensive move at the time because the effective range of his 20 mm. cannons far out-reached the 303 calibre machine guns in our turrets. This constant weaving continued for some time, with the shells of the faster Messerschmitt successfully raking our fuselage several times. It was brutally hard, tiring work, throwing my heavy charge around while still trying to maintain an average directional heading.

The starboard outer Merlin engine suddenly erupted in flames, as several cannon shells tore through it.

"Feather the starboard outer!" I yelled to the flight engineer, seated at my right shoulder.

He was holding on grimly as we climbed, turned, and dived in our attempts to evade disaster. Ken successfully shut down the stricken outer engine, then operated the No.4 Graviner fire extinguisher button. The propeller feathering system must have been damaged by the shells from the night fighter because it continued to windmill. This, added to the mangled condition of the engine cowlings, created enormous drag, which tended to create a yaw to the right and dropped our airspeed by about 20 miles per hour. I corrected for the yaw with left rudder and used the rudder trim wheel to relieve the load on my foot. Out on the wing the fire flickered for a few anxious minutes and then to everyone's relief, eventually died out.

"Rear gunner to Pilot. I've lost the Jerry, can't see him anywhere. I think I might have hit him, or he's run out of ammo."

"Good Show! But keep your eyes peeled. He might still be hanging around out there."

Gratefully I relaxed a little, realising I was bathed in perspiration which was rapidly cooling on my skin, due to the number of holes and draughts throughout the airframe. My arms and legs felt like lead and I was desperately tired. Flight Sergeant Dusty Miller, our Bomb Aimer who had been lying prone in the nose, now informed me that he had been violently airsick during our aerial ballet but thought he'd be O.K. when needed, close to the target area. He had of necessity, removed his oxygen mask before throwing up in order to keep his mask clean, and to prevent his microphone from being rendered unserviceable. He was a little bit woozy from lack of oxygen, but recovering well. The smell of vomit mingled with the acrid stench of expended ammunition fumes. I eased my oxygen mask to a more comfortable fit on my face and tried to breathe normally. Slight adjustments to the power output of the three remaining engines helped to balance the aircraft better. Thinking of our E.T.A. over the target area, I queried the Navigator.

"Bill, do you think we could cut a corner on this leg to intercept the final run to Berlin and make up for lost time?" He quickly answered yes, and proceeded to do so.

"Nav to Pilot, steer new heading, 120 degrees."

"Roger turning to steer 120 now."

The previously solid cloud cover below us was showing signs of breaking up, and the bomb aimer managed to catch sight of a good ground reference corroborating the navigator's Gee Fixes, and the radio op's loop bearings. It was good to know that we had been holding to the flight plan until now.

"Can you give us an E.T.A. for the target area?"

"Standby one Skipper."

Minutes later I was advised that we were now only thirty five minutes from Berlin. Already, ahead and below on the far horizon, we could discern a fiery glow and noted coloured flares dropped by the Path Finders to guide the stream of bombers. In the next few minutes I saw two more Lancasters spiral earthwards with fire licking hungrily along their fuselage and wings, but felt somewhat relieved when parachutes blossomed around them, clearly visible against the lit up sky. Were they victims of night fighters, or ack-ack? It was impossible to tell. Grimly I focused upon maintaining height and heading. Dusty Miller in the nose ahead of, and below my feet, was now ready to actively participate in his role. We were about to enter the most critical part of the operation, just two minutes behind schedule.

At the I.P. (Identification Point) easily seen in the glow of ground fires, plus moonlight on the waters of the Havel River and Wansee Lake, we began the most dangerous several minutes of the entire flight. To ensure bombing accuracy we had to maintain level flight. No corkscrewing. No violent aerial manouvers. No evasive action, just two minutes or so of straight and level flight, the huge bomb door gaping open. Any red hot shrapnel from exploding "ack-ack" shells, or one lucky cannon shot from an enemy fighter, and we could be blown to smithereens.

"Bomb aimer to Pilot, left, left, steady, ste-ady. Right a little. Steady—bombs gone!"

There was really little need to add the latter as the Lancaster, relieved of its bomb burden, always reared skyward for four or even five hundred feet. Even at this moment I had to try and keep the aircraft wings level so that the delayed action camera in the belly could take pictures of the exploding bombs, and confirm that we had been over the proper target area.

"Bomb doors closed. Let's get out of here Skipper."

It was then that the Flight Engineer informed me that we had lost quite a lot of fuel when No.4 engine had been clobbered.

"I've closed down the ruptured line and opened the balance cocks, but I think we'll have to fly a more direct route home."

I acknowledged and looked down at the blazing city. Should I feel nauseated, or guilty for the action we had just accomplished? No! This was war at its worst, or best. We were trying to defeat a fearsome enemy who was destroying lives and property. With a sweeping glance around for other kites, I banked steeply away and dove several thousands of feet to build up our departure speed, confusing the gunners manning their Flak stations.

"Navigator to Pilot, steer due west 270 degrees."

"Roger Navigator, steering 270 degrees and descending. Everyone keep a sharp lookout. We're not out of the woods yet."

Other gaggles of aircraft were due over the target area to drop bombs any minute now. We could easily become casualties of our own bombardment. We were also entering an area where it was common practice for the Jerry fighters to gather in hope of shooting down crews who were starting to relax somewhat, after the nerve wracking run up to target. It was like a scene out of Dante's Inferno. Below and all around, the city of Berlin was ablaze. The Bomb Aimer remarked to me that he could easily spot the explosions of the 4,000 pound Cookies as they struck ground. The 500 pounders, by comparison, were mere hiccups. Even the sky seemed aflame and the hot air rising caused turbulence, tossing F for Freddie around like a cork. Anti-aircraft fire was horrendous but I could now vary my heights and heading at will, in an effort to offset the radar guns' calculations.

"Pilot to Navigator—let's try to avoid the Hamburg and Bremen areas on this leg, they'll both be heavily defended."

"Nav, Roger, I have already plotted a course between them."

Just as we were sort of relaxing and feeling quite upbeat, the rear gunner announced that he thought he had seen another aircraft astern and slightly below.

"Couldn't identify it yet Skipper, but it might have been a Junkers 88."

Knowing that the twin-engined JU 88 night fighter was fitted with excellent radar and favoured attacks from astern and underneath, having a 20mm. cannon which angled upwards, we were all on our toes. I began a weaving pattern trying to avoid a constant heading. This also allowed Jock in the rear turret a chance to scan the suspect area below. Several minutes went by and just as expected, there was a cry from the rear turret,

"Corkscrew left NOW!!"

Once again we were in a struggle for survival, steep climbing turn to the left about ninety degrees. Nose up until the speed dropped dangerously close to stalling, and then a nose down dive, turning to the right with the wing pointing sharply earthward. Tracer fire arced from the night fighter and was returned from the mid upper and rear turrets in long bursts. Freddie took some more hits, and then our upper gun turret fell silent.

Jiggsy Giles's turret had taken a direct hit. When there was no answer to my request for him to report, I transmitted,

"Pilot to Bomb Aimer, can you check up on Jiggsy? See if he's alright and, if necessary, get the wireless op to take over the mid upper."

"Roger Skip, will do."

The front guns were of little use at this time because night fighters seldom made head on attacks. The bomb aimer worked his way aft, wriggling beneath the Flight Engineers seat. The Wireless Operator and Bomb Aimer lowered the gunner to a spot near the main spar and sadly informed me,

"Poor Jiggsy has had it. His turret is a shambles and totally U/S."

As quickly as the engagement had begun, it ended. Again we had been spared, but poor Sergeant Giles had sacrificed his life.

As we traversed west I noticed that the broken cloud cover was thickening up and quickly sought a level where we could take advantage of this, in an effort to avoid further confrontations with the Luftwaffe. Of course the clouds were no protection from the radar controlled anti-aircraft fire which often probed the sky around us. One close burst puckered the skin of the Lanc in several places and pieces of hot shrapnel shattered my instrument panel, knocking out the artificial horizon, the airspeed indicator, and my directional gyro compass. I had to slip my goggles down over my eyes to shield them from watering in the cold air. We were now scudding through turbulent cumulus clouds, relying purely upon the most basic instrumentation for control of attitude and altitude

Back to needle, ball, and airspeed, as my old instructor used to say. Only I didn't have an airspeed indicator. The V.S.I. (Vertical Speed Indicator) always tended to be too responsive and overreact, showing a higher rate of climb or descent initially, than was the true case. The altimeter, in direct contrast, was always slow to register changes in altitude. It was a tricky sort of balancing act to follow when all external references to the natural horizon were obscured. If the Lancaster's wings were not level, or if the nose pitched up and down even slightly, my magnetic compass would also tender false readings. I concentrated grimly on maintaining level flight, advising the navigator to monitor his own dual instruments. He had airspeed, altimeter, and directional gyro compass readings close to his right shoulder.

"Let me know if you think we are getting into any unusual attitude until we are clear of clouds," I advised him.

With much of the navigator's special navigational gee equipment damaged, and the radio equipment also "Gone for a Burton," it was simply a case of dead reckoning from this point on. Bill Taylor, having now bandaged a slight wound on the forehead, was busy trying to figure out our ground speed, distance travelled, and an E.T.A. for crossing the coast outbound. To aid him in this I requested that the Bomb Aimer return to his position in the nose compartment. Armed with a map he could assist by trying to observe any features on the ground he could recognise. Luck held with us and over an hour later, as the cloud began to break up, we recognised the watery but distinctive shape of the Zuyder Zee slide below the nose. We began to feel that we might make it after all. The wind-milling propeller of number 4 engine had finally ground to a halt, probably overheated due to lack of lubrication. This greatly reduced the drag and allowed our airspeed to creep up a little, to around 190 m.p.h.

The rest of the journey across the North Sea was uneventful. No other aircraft were sighted, nor did we have communication with anyone. It seemed like we had the entire sky to ourselves. We finally crossed the English coast at easily recognisable Spurn Head and the estuary of the Humber River. A few degrees change of heading then the very welcome sight of a red pundit flashing two recognition letters in Morse Code. These recognition beacons were always positioned several miles from the parent airfield but, as a rule, not always illuminated. Someone else must have recently requested it. Gratefully I banked around, onto the memorised heading for base. Unable to call the tower, I had the Wireless operator fire a two star Verey cartridge, using the proper colours of the day to indicate that we were friendly and requesting permission to land. A Green Aldis lamp signalled from the tower, and runway lights flickered on to orient me. I lined up and warned the crew to standby for landing. With Ken the engineer monitoring the power settings and Bill the navigator calling out the airspeed readings from his instrument cluster I requested,

"Under-carriage" down half flap, prop's fully fine, please Ken."

He answered shortly,

"Gear down Skipper, 3 green lights, flaps half, pitch full fine."

When the landing lights on the leading edge of the wing illuminated the trees about 200 feet below it was,

"Full flap please Ken."

With my navigator calling out the airspeed the rest of the way, we were crossing the perimeter track. I eased back on the power and raised the nose a touch. Early on a Friday morning, some eight hours and fifty minutes after take-off, there was a slight squeal of tyres as F for Freddie's wheels kissed the runway in a greaser of a landing. We were down. We were home again. With throttles closed, we gradually lost speed and braked to a walking pace. I turned off at the intersection of two runways and taxied back to dispersal, finally closing down all systems. I'd never felt more tired, or so relieved in my life. Another "OP" safely behind us.

An ambulance appeared and the body of our friend, Sergeant Giles, was reverently removed. In the cold grey light of false dawn, we six survivors gathered around a waiting truck which was to take us back to the crew debriefing room. I walked around my machine and wondered at the amount of punishment it could take and still carry us back home. As was my habit, I decided to walk back across the aerodrome on my own, leaving the rest of the crew to take care of the equipment, parachutes etc. This was my way of slowing down from the hectic pace we experienced in flight, to the more mundane and normal tempo of life on the ground. The crew had become accustomed to this idiosyncrasy. They too were glad that F for Freddie had brought them home safely.

On one such previous walk across the airfield in the early dawn, I had taken a shortcut through a familiar copse of trees and stopped when I heard the melodious warbling of a lone Nightingale, singing its heart out. It had been a brief moment of sheer pleasure as good as any concert, which did a lot to restore my mental equilibrium and soothe my nerves. I certainly would have enjoyed a repeat performance right then. Later in the debriefing room, we learned that the squadron had lost five Lancasters that night. A total of 35 men lost, not counting the casualties who made it home. We were also surprised to hear that Flight Sergeant Ned Sparks, the captain of Q for Queenie, had simply walked away from his aircraft and crew at the dispersal point. When he couldn't be found, a reserve Captain had been hurriedly summoned to take over the pilotless crew. They were reported among the missing, and had been observed going down in flames. When found later in his room having suffered an attack of nerves, Flight Sergeant Sparks was officially charged with L.M.F. (Lack of Moral Fibre) and within 24 hours he was posted away from the station in disgrace.

We now knew of course, that one day soon we would be on parade for the occasion of Flight Sergeant Giles's funeral. His promotion had just come through, and I was faced with the sorrowful task of writing a letter of condolence to his family. It was inevitable that these relatives would probably ask if they could view the remains of their loved one before the internment and, like many others before them, this request would be denied. "Just try to remember him as you last saw him," would be the gentle advice. All too often the remains, or what was left of them, would not make comfortable viewing. Sand would be added to make up the proper weight in each casket.

By this time most aircrew members had developed a sort of gallows humour which many people failed to recognise, or thought of as bizarre. In fact it was a sort of release mechanism which helped to alleviate the pain and kept us sane. We used expressions like, "He's bought the farm," or, "He's gone for a Burton," or, "Got the Chop," or, "The Big Hangar in the Sky will be full tonight," and other equally inane remarks when friends died. How else could we face these nightly forays? Of course we did it in the staunch belief that it would never happen to us. I had apparently developed a habit of saying, "See you when I get back," without being aware of doing so. I always felt sure that I would return, somehow.

****

Footnote:

Later, word had it that Ned Sparks had been reduced to Aircraftman Second Class (A.C.2) or Erk, the lowest rank in the Royal Air Force. He was still entitled to wear his pilots brevet, although doing so would expose him to ridicule or worse, as one who had failed in his duty. He'd been relegated to kitchen duties at another non-flying station. That seemed heartless to those of us who fully understood the strain imposed upon nerves by sustained exposure to danger. It could easily have happened to any of us especially the tail gunners, whose life expectancy in combat was often limited to mission totals counted in single digits. Similar to the "Shell Shock" diagnosis of World War I, this was something that should have been recognised and treated much earlier, in a more understanding manner, by those in Command. Sometime later, after the war had ended, LMF was recognised as an illness, and the slur on brave men who had been accused of cowardice later removed. This was of course too late to assuage the pain and bad feeling that had been engendered against some very gallant airmen who had faced the danger just one time too many.

A NIGHTINGALE'S GIFT

The airfield lay still in the pearly dawn.

All engines were silent, the airmen all gone.

As was my habit, I'd climbed from my plane,

To walk in solitude, upon Earthly domain.

In the breathless hush, as the sky turned pale,

There came the sweet song of a nightingale.

Fluting clear notes of sheer crystal beauty

Danced through the air, with absolute purity.

Lilting cadenzas, and trilling notes high,

Flowed out of its throat and into the sky,

To melt in my ear, and hold me spellbound.

I felt I was standing on hallowed ground.

How long I stood there, I really can't say.

Just found it impossible to go on my way,

As long as that concert, so freely given,

Wafted out of the trees, and drifted to Heaven.

It was a short spell of such sweet serenity,

An oasis of peace and utter tranquility,

Where I bathed my soul in simple delight

Until the last note. The melody ended,

And silence once more upon me, descended,

As daybreak heralded the end of the night.

Many times after, when ending a mission,

I retraced my steps, and oft' paused to listen,

In vain, for the glorious song of the mite,

Whose warbling entranced me that one magic night.

Tho' ne'er heard again, I still treasure its gift,

A moment of pleasure, that gave my heart lift.

I'll always remember that sweet wooded vale,

And the wondrous enchantment of one Nightingale.

MEMORIES OF THE BERLIN AIRLIFT

I have been asked the question, was the Berlin Airlift merely a chess move in a political game of chance, or truly a great humanitarian endeavor? The answer is a little more complicated than it may appear at first. Surely the Russians, when they closed all surface approaches to that city, fully expected their former allies (Britain, France, and the United States) to quietly fold up their tents and sneak away into the night, retreating entirely from the Russian zone of occupation in Germany. They figured that the Allies would not risk going to war against the mighty Russian Bear to defend a beaten enemy who had cost them dearly in men, material, and almost all resources for close on six years.

They were wrong.

Although some clever individual had foreseen the requirement for several air corridors leading into the divided city, no one really thought that a sustained airlift could possibly provide the massive amounts of food and other necessities to keep Berlin functioning. The Berliners themselves were of the same opinion, that the British, French, and American occupation forces would only make a token gesture before admitting defeat and succumbing to the Russian threat. Even the Allied

Triumverate suspected that an airlift, besides being prohibitively expensive, might possibly be beyond their capabilities, but were determined to give it a good try.

At first, the daily tonnage of supplies flown into the beleaguered city seemed to verify those pessimistic forecasts. The sheer logistics of gathering the assorted supplies, and providing the crews and aircraft to deliver the goods seemed overwhelming, but gradually the impetus grew as more planes were diverted from sources around the old theatres of war. There was also a noticeable change in everyone's attitude. As the tempo increased, so did the enthusiasm of everyone employed in this huge task.

When the airlift first started, known variously as Operation Plane Fare or Operation Victuals, most crews were just following orders and indeed, many servicemen felt reluctant to provide succor to the former enemy, but complied anyway. I was one so inclined, but gradually my mindset underwent a subtle change. Instead of flying on missions of destruction, we aircrews were now employed on missions of reconstruction and mercy. The enthusiasm spread rapidly and soon everyone became involved in trying to increase the efficiency of the operation. Citizens of Berlin, joyfully recognising these efforts, also contributed greatly with many more labourers and constructive improvements to Templehoff and Gatow Airfields. A third airfield was then being built at Tegel, to help reduce aerial congestion.

As this new found trust developed, there was more awareness on the civilian side of the true commitment being made by the Allies. Realisation as to their way of life, compared to that in the Russian sector of Berlin, could be considered as the first weakening of the Communist doctrine. Later, when it came to actually voting in the free elections, this became very evident and the Red Party suffered. Little did the Russians realise that, by creating the necessity for an air lift in 1948, they were providing their former allies with a perfect training tool for future use in any similar emergency around the world. All personnel concerned with the logistics of such an operation were being given ample opportunity to perfect their knowledge.

The greatest benefits were experienced by the crews of each aircraft in flight. Although instrument-flying, and so-called blind landing approaches had been practiced and improved upon over a number of years, the bad weather flying we experienced during the sustained airlift proved invaluable. In wartime, many operations had to be scrubbed because of bad weather. Too many aircraft had crashed, and too many airmen had died when attempting to land in conditions of poor visibility. I can recall feeling quite relaxed and comfortable a number of times, conducting a completely blind instrument approach to Gatow. The reassuring voice of the experienced G.C.A. operator, talking me down and telling me exactly where I was in relation to the runway, was a soothing experience. More than once, just as I was informed that I was now crossing the perimeter fence, my first glimpse of the runway lights and actual sight of the ground since take-off popped into view a second or so before my wheels kissed the tarmac. Such blind approaches to landing would have been frowned upon, or prohibited in earlier days.

Had the Allies indeed been forced to go up against the Russians at this time, we would all have been at a higher standard of preparation and capability because of the all-round experiences gained during the logistical miracle. Fortunately, we were never put to that test. Of course in such an endeavor there were accidents, and fatalities. Such is the price that had to be paid, but we felt that the men and women who perished during this epic operation did not die in vain.

There were also numerous and humourous little episodes, where valuable lessons were learned. One very dark and snowy evening in March, 1949, as I was taxiing a heavily-laden Dakota DC3 toward the take-off point at Lubeck Airport, I was asked by the air-traffic controller if I could stop and take a passenger aboard for Berlin.

Assured that he was authorised, I agreed and stopped. My radio-operator hurried to the rear, opened the door, and set the steps in place for the passenger to climb aboard. Our cargo that night was several hundred bags of coal. As the radio man returned, I asked him who and where the passenger was.

"Oh, he's some high-ranking officer, and he's sitting on one of the coal bags near the door," was his reply.

"He will freeze back there. Go and tell him that there's a place up here beside me."

A few moments later a tall, distinguished gentleman appeared, lowered himself into the co-pilot's seat, and strapped in. He then introduced himself as General McLean, a staff-officer of the Canadian Army. His previous immaculate greatcoat, known as a Service Pink, was liberally smeared with long black smudges where he had come into contact with our inanimate, dark cargo.

Once we were airborne and the undercarriage fully retracted, we were immediately swallowed by solid cloud cover, severe icing conditions prevailing. I was kept busy for a while, employing all de-icing rituals. There were several loud bangs as the ice centrifuged off the propellers and slammed against the metal fuselage. The rubber boots on the leading edges of the wings were similarly busy preventing any ice build-up. For about ninety minutes, all the way to Berlin, we were cocooned in this environment. The runway at Gatow did not appear in view until we were less than one hundred feet above ground and a quarter of a mile from touchdown. When leaving us, General McLean expressed his sincere admiration for the crew's efficiency, and thanked us graciously as he exited the aircraft. He seemed totally unconcerned about the state of his pink and black greatcoat, and was assuredly happy to have both feet back upon terra firma…the more firma…the less terror!

One morning at Gatow the weather was foul, with freezing rain and clouds so low, tall pedestrians were ducking. Even the most experienced pilots were prudently awaiting some signs of improvement and enjoying the brief respite. Air Vice-Marshall Bennet, of Pathfinder fame, was amongst those who were chafing at the bit. He had arrived in a new Avro Tudor Airliner and the moment he was given clearance to chance his luck he dashed off, performed a less-than-accurate pre-flight check, and was soon lifting off from the runway. Once airborne, he realized that the elevator locks of the Tudor were still in place and he did not have full control in the pitching plane. With no other craft in the immediate vicinity (a rare occasion) he managed to fly around the circuit and land safely. A crew member dashed out and removed the offending wooden locks, allowing Vice-

Marshall Bennet to depart without revealing his red face. All's well that ends well!

On another occasion at Gatow, it was dusk as I taxied out to return to Lubeck for another load. I was behind a Yankee Skymaster at the time, and listened as he asked for permission to line up on the runway for take-off. The traffic-controller answered,

"Negative. Hold your position. We have one aircraft on short finals."

Naturally we looked out to our right and spied a Royal Air Force York (Civil version of a Lancaster) emerging from the low overcast with undercarriage fully down, flaps extended like huge barn doors, and his navigation lights blinking. Visibility was reduced by the light snow falling and a blanket of snow covered the ground. Over the button of the runway the York pilot rounded-out a bit high, obviously misjudged his height slightly, and then dropped to the ground with a bone-shaking jar. As the machine bounced back into the air, the pilot gunned all four throttles in an effort to maintain control. Then he powered-off again. Once more the York made heavy ground contact, which caused a second bounce as it proceeded down the runway. The Skymaster pilot ahead of me then announced over his microphone,

"Ah, say there Kangaroo, when yuh gits to the end o' the runway, hop to the laift!"

Of course the American crews were notorious for their habit of being rather verbose on the common radio frequencies. The R.A.F. rather frowned upon anyone "hogging the air" as they called it. Strict use of terminology and brevity expected at all times! I knew pilots to be hauled before the Senior Air Traffic Controller (SATCO) for minor breaches of this rule. Of course profane language was verboten. There were times though, when one appreciated the witticisms overheard in passing.

For instance, amongst the various and many types of aircraft used on the airlift was a twin-engined machine known as the Bristol Wayfarer. This was strictly a cargo freighter, owned by Silver City's Airways based at Lympene, in Kent. Used as a car ferry across the English Channel, it was a high-winged monoplane, with two powerful radial engines driving two props. It had a fixed undercarriage, a slab-sided fuselage, and two huge clam doors in the nose to expedite the loading and unloading of its vehicular cargo. The crew compartments sat high above the nose. It would never win any beauty competition, but the pilots who flew it commented on its reliability. Descending toward Gatow one day, clear of cloud, I saw this freighter plodding along. An American pilot who had been back chatting with one of his airborne buddies, also spotted the old-fashioned Wayfarer and uttered these words,

"Hey Hank, these Limeys are sure throwing everything into this operation. I just passed the Mayflower!"

One of the compulsory reporting points on the way into Berlin was a radio beacon called (phonetic spelling) "Frownow." R.A.F. pilots reporting there would state,

"Air Force 621 overhead Frownow at 5 point 5 I.F.R. Load, coal," thus giving tower control information as to aircraft call sign 621, position Frownow, elevation 5,500 feet, in cloud, Instrument Flight Rules.

The controller would then know where to direct the aircraft after landing, for unloading. Pilots would be instructed to change radio frequency for their final approach, and listen for their G.C.A. talk-down.

An American pilot reporting over the same beacon would be likely to say,

"Triple-deuce, over Frownow, in the soup, toting black stuff!"

Depending on the load carried, these conversations could sometimes be very colourful. Pilots like me, flying the venerable old Dakotas, swear that they are one of the finest, most reliable, and forgiving aircraft ever built. I can give you an example. Prior to boarding my aircraft at Lubeck one very cold wet evening, my crew and I were soaked by the freezing rain coating everything. Completing our external pre-flight checks as quickly as possible, in order to avoid further soaking, I checked that our cargo was securely strapped down to the strong points on the floor. It didn't look like much of a load, just several large steel wheel-gear-machinery parts, and didn't take up much space. I removed my heavy coat and entered the flight deck to commence engine start without consulting the freight manifest. That was a mistake.

Before long we were cleared to taxi and given permission to take off. There was a few inches of slushy snow on the runway, as we commenced our take-off run. Acceleration seemed a bit sluggish, even with full power, which I attributed to the state of the runway. However, as the length of our run increased and the amount of runway remaining diminished to a dangerous degree, I realized that my machine was not yet ready to fly. I hurriedly selected 15 degrees of flap to provide more lift, and the willing old Dakota staggered off with literally just a few yards to spare. As our airspeed was barely above the stall, I quickly retracted the undercarriage to lessen the drag, and we slowly climbed away. Too close for comfort. Safely on the ground again at Gatow, I found that we had carried a load intended for the four-engined York transport.

The Russians were well-known for conducting nuisance tactics to disrupt the flow of Allied aircraft in the air corridors we used. On one such occasion, I noticed a Yak fighter heading toward me at high speed. His intention was obviously to make me veer away and, perhaps slip out of the safely authorized corridor to become fair game for them. I assured myself that the crazy Russian pilot would not kill himself by crashing into me, so I merely lowered my seat a couple of notches to concentrate on straight and level instrument flight. Once foiled of any reaction on my part, the Russian departed to seek other prey.

In May of 1999, my wife and I attended a reunion celebration in Berlin to mark the 50th anniversary of the ending of the Berlin Airlift. We travelled free as guests of the city, and we were treated royally by everyone we met. It was most enjoyable to meet up with so many old friends, and the organisers worked very hard to ensure that every visiting veteran received the honours happy Berliners felt were due to them. It felt strange to have hausfraus come up to us in the streets to shake our hands, or to give us heartfelt hugs of affection. The closing ceremonies were also quite moving. I was proud to be the bearer of the Canadian Flag as we marched into the Olympic Stadium, where the Mayor of Berlin personally pinned medals on the chests of all participants. All in all, it was a fitting memory of days gone by and a job well-handled. The Allies had shown the world that it could be done. During that year, over six and-a-half million tons of supplies were air-lifted into the city.

INCIDENT IN BURMA

During the last few months of World War II, I was stationed in Burma as a pilot attached to number 62 Squadron of the Royal Air Force. This squadron was equipped with the venerable Douglas DC3 aircraft, popularly known in the service as the Dakota or simply, the Dak. Our major tasks were to support the army units which were operating in the jungle warfare against the Japanese. Our airborne operations consisted of dropping supplies of food, munitions, and any other necessities to the foot soldiers struggling through extremely difficult and unfriendly terrain. Much of the heavier supplies were of course dropped by parachute to alleviate damage. But a great deal of urgently-needed supplies were also dropped at extremely low altitudes, using a system which became known as the S.E.A.C Drop. This was the abbreviation for South East Asia Command.

Supplies of food such as rice, flour, wheat etc. were usually loaded quite loosely into jute sacks which could stand the harsh punishment of striking the ground when dropped from an aircraft. Each was put into a second sack to reduce any possible spills, total weight around 86 pounds. Prior to reaching the operational D.Z. (Dropping Zone) the crew of the Dakotas would pile up as many sacks as possible, close to the large open door in the rear-left end of the fuselage, and await a signal from the pilot as he adjusted their approach. Having identified the zone to be friendly, the pilot would then descend almost to tree-top height and reduce speed to a safe minimum. When judged to be within the correct parameters for dropping, the pilot would operate a switch which flashed a green light and rang a bell over the doorway. That was the signal to heave as many of the sacks overboard as quickly as possible. This action would cease as soon as the pilot cut the green light and bell signal off. In many selected D.Z.'s, an overshoot of any of the dropped supplies could mean a total loss of same, and could possibly help to feed the enemy.

This circling and dropping procedure would be repeated until all of the necessary supplies had been successfully jettisoned to the troops. With practice, many of the Dakota crews became quite expert at placing the freely-dropped supplies right onto the coloured markers identifying the target area. I heard a Scotsman brag one day that, if it was a football field down below, he could score a goal anytime he wished to. For us, the pilots of these venerable aircraft, low-slow flying became a routinely normal procedure that was considered quite safe because of the Dakota's well-deserved reputation and reliability. It could fly safely, within certain weight limits, even with one engine shut down.

When the Japanese forces eventually surrendered and the war was declared officially over, many of the servicemen stationed in the Far East expected to be repatriated home very quickly but, for many of us in Burma, that expectation was very quickly squashed. By this time our squadron was stationed near the capital city of Rangoon, at an air base called Mingaladon. Mingaladon had one long runway, surfaced with pierced steel planking. From our parking site we could see the golden spires of the beautiful Shwe Dagon Pagoda, a well-known landmark. It was visible for miles in any direction when approaching Rangoon from the air, gleaming in the tropical sun.

Before long we discovered that the government was planning to use the Dakota squadrons to help the Burmese survive an imminent famine. The indigenous people who lived in the mountains, and more remote areas of this jungle-draped country, were in danger of starvation because they had destroyed their rice crops in an effort to deny the Japanese invaders this easy supply of food. The people had even burned their supply of seed, which was usually held to create the next season's crop. Many of these brave folk had suffered greatly at the hands of the Japanese because of these actions, hundreds executed in reprisal. Now the task of feeding them became our operational priority.

Instead of searching for troop locations in the jungle, we were now given local maps of districts to be covered, along with instructions as to how many tons of rice, or other grains, were to be delivered. 62 Squadron rose to the occasion proudly. For the next few months I found myself operating deep into the mountainous areas of north-eastern Burma, often quite close to the border with China. Our task was to locate small settlements, far removed from civilization, and free-drop the life-saving supplies. This was an exhilarating and very rewarding experience, giving me great personal satisfaction because of its humanitarianism, and because it afforded me the opportunity to travel widely over the ancient and beautiful country.

Back at Mingaladon, life resorted to a more humdrum existence. We were accommodated in large tents holding four-man crews, awaiting news of when we might be repatriated to Britain. A number of people came up with the expression, "Roll on the boat," and started marking their calendars. Most of the troops were, of course, boarding troop transport ships for that welcome journey home. In a wired-off section of our military camp there was a large E.P.I. tent, which contained a supply of American "K" Rations. These contained quite a variety of desirable articles such as chocolate, chewing gum, cigarettes, cans of soup, packets of cheese, and even sheets of toilet paper. Although this section was strictly "Out of Bounds" to us, I used to crawl under the fence quite regularly, liberating a few boxes of these luxury rations to share with my crew. To heat soup or boil water for tea, my crew members used to scoop a shallow declivity in the soil just a few feet from the entrance to our tent, and then light a small fire. Each day we removed the old ashes, and replaced kindling twigs for the next cooking session.

One day my navigator, Bill Taylor, was taking care of the fire and attending to tea-making when, to his surprise, the ground sort of opened up, dropping dirt, fire, and tea-making equipment into this newly-formed hole. He stepped back and called us out of the tent, showing us what had happened. As we bent over the scene, there was a sudden "whoosh" from below, and fireworks of some sort soared into the sky! This was quickly followed by a few more star shells which we could identify as coloured signal cartridges, such as those fired from a Verey pistol. After describing graceful parabolic arcs, these colourful flares descended to the ground within the lines of domestic tents, quickly setting some on fire. Within minutes, a general fire alarm was sounded, and everyone was called upon to fight the blaze. One of the first casualties was the huge canvassed area of the Officer's Mess. The fire spread with amazing speed, quickly consuming the few dry bushes which dotted the perimeter of the domestic site. I found myself standing side by side with the Station Commander, both of us perspiring freely as we swiped at the flames with a piece of tarpaulin. I noticed that a tractor had been brought into action nearby, in an attempt to make an earthen fire break which should help contain the conflagration. But this merely revealed another, greater problem for all concerned.

As the tractor's very large blade scraped away soil, it suddenly struck metal and exposed a number of huge steel casings. These were, in fact, highly explosive bombs of the type usually dropped from Japanese aircraft. Unknown to us, we had been living for months, right on top of a concealed bomb dump! All attention was then diverted to preventing the fire from spreading to this area, only a short distance from the line of parked Dakota aircraft. A few minutes later the station tannoy announced that all captains of aircraft must report to the operations section immediately. We were quickly briefed to move the parked Dakotas away from the endangered area. Since parking areas were rather scarce elsewhere, and we couldn't block the active runway, it was decided that some of the Daks would have to be flown to another base until the emergency was over, so we headed out toward the west coast, landing at Mhawbi for 24 hours. I was relieved on returning to find our own crew tent had been spared from the inferno, and our belongings were still intact. It was shortly after this I had the opportunity to fly to Changi airport, on Singapore Island, taking part in another and even more welcome mercy flight. We were to ferry ex-POWs back to England. For that reason I was not present when, and if, any enquiry was held concerning the fire at Mingaladon, so I have no further information to offer on that score. Suffice to say, it certainly was a memorable event and quite innocently caused. Fortunately no one was injured in any way by this except perhaps, by the loss of some personal belongings.

FOLLOW UP TO BURMA INCIDENT

Flying ex-POWs back from Changi to Britain, many of my passengers were in such poor physical condition that they were stretcher cases, having been badly beaten and tortured by their Japanese guards. They were attended to by nurses during the entire long series of hops between Singapore and Lyneham, near Swindon, in Wiltshire. On my periodic walks through the cabin I took the opportunity to talk with some of these servicemen, all naturally pleased to be going home after years of war, deprivation, and separation from loved ones. Plasma bottles and other medical tubing were attached to some of the patients who were mostly skin and bone. I had my wireless operator radio ahead to each intended landing field, advising the authorities there of the nature of my mercy flight. Thus, we were always met by well-prepared staff who would organise meals, medications, and suitable accommodation for everyone if we decided upon an overnight stay.

Finally, the green fields of England hove into view, and I was touched to see tears rolling down many of the emaciated faces as they disembarked, or were carried from the Dakota fuselage. The Customs and Immigration people were extremely considerate, and soon everyone was on their way to be processed for military hospitals, or other receiving units. I was fortunate enough to wangle some furlough time and was soon home with my wife, an ex-W.A.A.F. This entire trip, the longest I had ever made, had left quite an impression on me, but it was soon relegated to the back of my mind as I continued my service career in other posting and flight duties.

Several years later, I was stationed at RAF Station North Luffenham, employed as a staff instructor, converting pupils onto the Dakotas of Transport Command. My wife and I were renting a small cottage in Seabrooke, between Hythe and Folkestone, in Kent, so I used to hitch-hike home on as many weekends as possible. Returning to camp on the Sunday evenings involved taking a train from Folkestone to Charing Cross railway station in London and then, using the Underground to rush across to King's Cross, catching a specific train to Peterborough. There was special transport laid on each Sunday for the approximately twenty mile drive to North Luffenham. Woe-betide anyone who missed that bus, because it meant a long, lonely hike, or an expensive taxi ride back to camp.

One Sunday evening I reached King's Cross station a few seconds late, just in time to see the rear end (caboose or guard van) of the train disappear as I raced through the platform gates. Naturally I finally reached Peterborough in a later train at some ungodly hour of Monday morning, to find there was no bus awaiting me. I started to walk, hoping to thumb a lift because I had to be on parade by 08:30 hours! Several miles later, without having seen a single automobile going in my direction, I was beginning to think that my luck had run out. It was then I heard a car approaching from behind.

"Please Lord, let this car stop," I prayed, hoping perhaps it would at least carry me just a few miles ahead, to the busy A1 main north-south highway. There was always a lot of traffic in each direction there. I stood in the centre of the roadway and gave the universal sign of the thumb. As the vehicle stopped my heart sank. It was in fact, a taxi. Poorly paid as we were in those days, I could scarcely afford such a luxury, but it looked like this was going to be the only way I could be in camp on time for the C.O's Parade, or I'd be in trouble, so I walked over to the car.

After speaking to the driver and stating my destination, which he had already guessed, I opened the rear passenger door to enter the cab. This lit up the roof courtesy light, and the driver turned to stare at me rather fixedly.

"I see you're a pilot," he said, then proceeded to ask me a series of questions, "Were you stationed in Burma at the end of the war? Did you fly Dakotas? Did you pilot a bunch of ex-POWs back to England, from Singapore?"

To these and every other question, my answer was in the affirmative. The driver then leaned over and asked to shake my hand, as he informed me that he had been one of the stretcher cases on that particular flight.

"I used to lay there on that stretcher and look into the crew compartment every time the flight deck door was open, and you spoke to me several times during the journey. I've always wanted to say thank you, but when I got out of the hospital, I didn't know how to reach you."

Naturally we were both very surprised by the amazing serendipity of our strange meeting. I for one, would never have recognised this man in a crowd. Last time I had seen him he was an emaciated, skeletal figure, encased in army blankets and being fussed over by the attending nurses. We spent the rest of the journey to the camp gates reminiscing about the flight and catching up on events of mutual interest.

When I was leaving the taxi at the guard house, the driver again shook my hand and refused to take any money from me. Instead, he presented me with one of his business cards and told me that I could call upon his services at any time, day or night, if I ever needed another lift. I thanked him and was very grateful for his offer. I never saw him again.

The Bible says, "Cast your bread upon the waters, and it shall be returned to you a thousand-fold."

Well, in this case it was true, and I was indeed a very fortunate man.

AIRCREW WALKS ON WATER!

Some years after the war ended, my crew and I were sent on a sea survival course, at R.A.F. Thorney Island, on the south coast of England. The course was to acquaint us with the latest ideas on survival equipment such as the newest rubber dinghies, and radio equipment etc. We were to be introduced to the air drop-able, motorized lifeboat which could be carried in the bomb bay of the Lancaster bomber, of wartime fame.

This was the sort of course the aircrew would welcome if dispatched there during the warm summer months (as most officers were) but, since we were all of non-commissioned rank, we were now to experience the rigours of this chilly Solent Estuary in mid-winter…brrrr! Lectures on the most up-to-date equipment were held in warm cozy classrooms. Here we familiarized ourselves on the newest radios and other assorted gimmicks, including the various rubber dinghies available for aircrew use. But that comfortable session soon came to an end.

One day we found ourselves shivering at the water's edge, wearing our usual battle dress and Mae Wests. The Estuary here was approximately one mile wide, with the far shore a dim low smudge on the horizon. I reckoned the river current was running at about five knots. Dinghies of various sizes and shapes were laid out for our inspection, assorted crews allotted to each type before heading into the grey and very cold-looking river. My crew was selected to man a "Q-type" canoe-shaped model. We slid it carefully offshore, trying desperately to avoid getting wet. Of course, that was the moment it began to rain!

In no time at all there were no fewer than ten yellow survival dinghies dotted well out on the water surface. A Royal Air Force Air Sea Rescue power boat was now present to act as our watch dog, in case of emergency. This powerful launch started circling whilst assorted crew members, armed with binoculars, kept a watchful eye on the shivering flock manipulating the life rafts. Using the abbreviated paddles, my crew and I were fighting a losing battle against the outflowing current, which quickly had us well out into the middle of the estuary. We started taking on some of the choppy waters and had to keep bailing it out. Between that and the rain, we were all soon soaked to the skin, and the moans began.

"Why are we here?"

We were feeling very sorry for ourselves and envying the crew in snug warmth of the launch, no doubt enjoying a nice hot cup of tea, or even something a bit more bracing! Gazing sideways at the far shoreline, I spotted a lone tree and after a few minutes, realised that we were holding a constant position, relative to it. There was a slight bow wave on the prow of our dinghy and that gave us the impression of movement, despite our now languid attempts at paddling.

The water was dark but I was now firmly convinced that we, ourselves, were stationary. I lowered my hand over the side and into the water up to my wrist. It confirmed that we were indeed aground upon a hidden sand bar, which felt quite firm to my touch. Deciding to give my crew a bit of a shock, I suddenly yelled,

"I've had enough of this! I'm heading for shore!"

With that I stood up and jumped overboard, the crew too stunned to stop me, and found myself standing just ankle-deep in the water. After overcoming their surprise, everyone followed my orders to get out. Three to a side we bent over, picked up the life raft, emptied the water out, and started to walk across the sand bar to deeper water on the other side. It was at this exact moment the senior officer aboard the safety boat turned his binoculars towards us...and his eyes bugged! There, about half a mile from shore, were six aircrew men walking upon the water, carrying their dinghy between them.

I was informed later that he took our Lord's name in vain and promised never to ever let alcohol pass his lips again. The Lancaster roared overhead and dropped the airborne lifeboat my crew were supposed to board, but that's another story.

ROYAL AIR FORCE FLY-PAST

Queen Elizabeth's 1953 Coronation (happily in tandem with an anniversary of the R.A.F.) made the powers that be decide to stage a formation fly-past over Buckingham Palace. The idea was to present an unbroken stream of as many different types of aircraft as possible, in a Royal Salute to the Queen.

This formation was to be led by a helicopter dangling a huge weighted R.A.F. Ensign suspended from the machine's cargo hook, and included basic piston and jet engine Trainers, Fighter-types, Bombers, and Transport aircraft, which all had different cruising speeds, as well as vastly different performance capabilities. The task of coordinating all these different types of aircraft into one continuous, unbroken stream over the Palace would of course involve split-second timing, using a number of assembly areas and rendezvous points for slotting the faster machines in behind the slower propeller-driven trainers, and keeping all running smoothly.

The man selected to organise this operation was Air Marshall, The Earl of Bandon, a very experienced pilot well-beloved by all who had served under him during his career. He had been a young fighter pilot and a, "bit of a lad," during the Battle of Britain, when his title was changed slightly by those who knew him well, to be popularly known as "The Abandoned Earl." Of course, few people ever addressed him as such to his face, but he was known to revel in his well-earned nickname.

The Air Marshall decided that the R.A.F. Station at Odiham, on the western outskirts of greater London, would be the ideal place to make the Centre of Operations for his rather formidable task, and duly arrived there to be met by the station's commanding officer, a Group Captain.

"I shall require an office, somewhere close to the Air Traffic Control building, and preferably with a good view of the airfield," stated the Earl, "I shall also require some large-scale maps of this district, as well as the central London area."

"Certainly Sir!" replied the Group Captain, "I think we have the perfect spot for you. The Station Warrant Officer has an office on the ground floor of the control tower, and his window looks out onto the airfield. I shall see that he vacates the office immediately. You will also have direct communication with the Air Traffic Controller on the floor above you."

In the meantime, a helicopter seconded from the Central Flying School, Helicopter Wing was already situated at Odiham, and earmarked to be the proud carrier of this huge R.A.F. Ensign Flag that would flutter suspended from the machine's cargo hook, at the head of the formation. The pilot of this machine was accompanied by a Flight Sergeant mechanic who was responsible for maintenance of the helicopter. Having discovered a minor problem with the cargo hook, it was decided that a spare part be ordered from home base to fix it.

"Not to worry, Sir," the Flight Sergeant said, "The Station Warrant Officer and I are old friends. We joined up as Boy Entrants at the same time. I'll just nip over to his office and use his phone to call base, and order the part we need."

With that he jauntily headed across the grassy strip and perimeter track toward the control tower, totally unaware of the pitfall that awaited him there. Reaching the door of the S.W.O. office, he swung it open without knocking and espied a figure with his back to the door, bending over a large map spread upon the desk. The blue barathea cloth of the uniform trousers was spread tightly over a rather plump derriere, similar in size and shape to his Warrant Officer friend's ample build. It presented a very tempting target.

Taking two swift strides into the room, the Flight Sergeant gave the trousered seat a resounding whack on the backside as he uttered,

"Hello you old bastard, do you mind if I use your phone?"

The startled individual began to straighten-up and turn around, leading this movement with an automatic reflex of rubbing the afflicted portion of his anatomy with his right hand. To the Flight Sergeant's horrified gaze, instead of the Royal Seal depicting Warrant Officer rank on the revealed tunic sleeve, there was a very large, dark ring, topped by another three, more narrow rings. He almost died of shock as he was further presented with the front view of a rather surprised Air Marshall, complete with Pilots Wings and several rows of multi-coloured ribbons denoting decorations, honours, and campaign medals.

Standing rigidly to attention, the poor embarrassed N.C.O. attempted to stammer out an apology.

I'm s...s...sorry, S..s..sir! I...I...I th...th...thought you were s...s...somebody else!"

Fully expecting to be exposed to the *very* senior officer's wrath...and visualizing arrest, with possible charges of Les Majeste or worse, he was surprised when the Air Marshall merely answered,

"I thought so, Flight Sergeant. There's the 'phone...make it snappy," before returning to his perusal of the spread map.

This of course was an indication as to the character of The Abandoned Earl. Instead of venting his anger on the younger man for his unwise, though unintentional attack on his dignity, he was quick to see the humour in the incident, realising he had been mistaken for the station Warrant Officer. He didn't even consider any reprisal in fact, in the Officers Mess later that same day, he took pleasure in recounting the incident with great gusto.

"You should have seen the look of horror on that Flight Sergeant's face," he laughed, "I'm sure that he expected to be shot at dawn."

His appreciative audience laughed too. Meantime in the Sergeants Mess, a Flight Sergeant (who shall remain anonymous) was also regaling an enraptured audience with his version of the event.

"Meet the only senior N.C.O. in the Royal Air Force who could walk up to an Air Marshall, give him a hearty smack on the arse, call him an old bastard...and get away with it!"

Of course that was said in jest, since everyone present was now well aware of the switch in office occupancy and the speaker's mistake in not identifying his target.

"If it had been any other officer rather than the Abandoned Earl, my son," offered a grizzled Warrant Officer, "you'd be in jail right now, awaiting court martial. Consider yourself lucky!"

Incidentally, because of the clever planning of the Air Marshall, the fly past over Buckingham Palace went perfectly, an unbroken stream of various aircraft in impressive review to celebrate Queen Elizabeth's Coronation.

SWIFT VENGEANCE IN CYPRUS

The following true story is about a brutal murder and the events immediately following, but the author craves the indulgence of his reader to allow a proper build-up of the circumstances. The locale is the island of Cyprus, 1956.

Cyprus, birthplace of the legendary Aphrodite, is a green jewel of an island nestled in azure waters at the eastern end of the Mediterranean Sea. Steeped to its marrow in history, this magic isle is blessed with a benevolent climate. Rich in copper, asbestos, and timber, it is also famous for its citrus fruits, vineyards, and olive groves. The population is mainly Greco/Turk, with the Greeks outnumbering the Turks by a ratio of approximately four to one. Under British Mandate, these two highly-volatile groups had learned to live in comparative harmony despite their differences in dress, religious beliefs, and ethnic backgrounds but, like water and oil in a jar, the longer they settled, the more pronounced and clearly delineated their differences became. No permanent blending appeared possible.

By the early 1950's open rifts sundered the fabric of society and the tension slowly mounted. With its own coffers sadly depleted and nearing bankruptcy, mainland Greece coveted possession of the island and its riches. Through the efforts of the Orthodox Church and Greek government, the Cypriot-Greek population was urged to strive for closer ties and actual unity with their Motherland. More and more often on this idyllic island, the blue and white flag of Greece was defiantly hoisted high above churches and government buildings, after the Union Jack of Britain had been torn down and trampled in the streets. Harassed "Tommies" found themselves the targets for stone-throwing villagers, and were hard-pressed to maintain their dignity or control. The hitherto pristine white walls of government buildings were daubed with huge blue painted slogans screaming, "Enosis!" meaning "Union!" with Greece. Doors and walls of mosques, hallowed to the Turkish residents, were also desecrated in like fashion.

Slowly the hate grew. Acts of vandalism became the order of the day and gave birth to a freedom movement, using the initial letters E.O.K.A., which joined the "Enosis" signs on walls all over the disturbed island. All too soon these so-called freedom fighters, armed with weapons smuggled in from the Greek mainland, resorted to murder in an all-out effort to further their aims.

Ambushes on lonely roads and farms became commonplace. Young British servicemen (many of them mere nineteen year-olds, serving their two year conscription period) were assassinated in the crowded and narrow streets of Nicosia, shot in the back by cowardly thugs who then quickly melted away before the area could be cordoned off. Many personal grudges with Turkish merchants were also settled in this fashion, tension rising. As a British garrison of long standing, it was an accepted practice to have the families of married service personnel accompany the men on their two-and-a-half-year tour of duty on Cyprus. These families contributed a great deal to the island's economy, but now some of them were to pay a terrible price.

Not content to wage their quarrel on a man-to-man basis, the E.O.K.A. terrorists turned their attention upon innocent women and children in a vicious campaign. Fire bombs razed family dwellings to the ground. Buses transporting children to schools were raked by machine gun fire, and deadly homemade bombs killed or maimed pregnant wives as they drove the family car on shopping errands. No one was safe. Despite strictly-enforced curfews, manned road blocks, and regular patrols, these terrorists seemed to be able to strike at will then just vanish like will o' the wisps.

In 1955 the newly-appointed Governor of Cyprus, Field Marshall Sir John Harding, was quick to size up the worsening situation. By his directives, the military strength of the beleaguered outpost was quickly and greatly increased. The colourful uniforms of the Special Air Services, Parachute Regiment, and assorted Highland Regiments were soon thronging into hastily enlarged barracks, and the tactical wings of the Royal Air Force were expanded by the arrival of extra fighter planes, transports, and the ever-useful helicopters. These last mentioned were to afford ground troops the tactical mobility which would turn the tide against the insurgents. Cyprus was now an island fortress bristling with weapons, but the guns did not point outwards. The enemy lay within.

Secret mountain trails used by terrorists and hitherto undetectable at ground level now became easily visible to the aerial watch dogs in slow-flying helicopters. One by one, lairs as remote as eagles' aeries were carefully pinpointed. Courier routes and food caches, as well as mail drops were uncovered to be plotted on Army maps, enabling ambushes to be laid with increasing success against the guerrillas. Troops airlifted to these remote areas by the versatile whirlybirds of the R.A.F. encircled the desperate E.O.K.A. converts in their hideouts and cut off their planned escape routes. Slowly the noose tightened. Terrorists previously used to hit and run tactics, disappearing chameleon-like into well-camouflaged caves after, found themselves outpaced and outnumbered by the tactical flexibility of the security forces. The Troodos and Kyrenia Mountains, scene of most of these actions, echoed more and more with staccato rattling of gunfire as opposing forces met head-on in desperate clashes where usually no quarter was given, expected, or asked for. The mere possession of arms or explosives, if discovered, merited the death sentence, so gunmen rarely surrendered.

One morning by sheerest of co-incidence of timing, I found myself drawn into a whirlwind chain of events immediately following the brutal murder of a British serviceman. But this time, the action was far-removed from high mountain peaks, centering around the Nicosia Airport instead. On this occasion vengeance recoiled upon the E.O.K.A. trigger men with a deadly swiftness that sealed their own doom. For this reason I have named the following true story:

WINGED NEMESIS

On the morning of May 16th 1956, Corporal "Paddy" Hale of the Royal Air Force was standing in the doorway of a radio shack situated near end of the main runway for the joint civil/military Nicosia Airport. It was pleasantly cool in the shade of the hut, as he watched perspiring groups from an R.A.F. Regiment toil to erect a massive barbed wire fence around the perimeter of the 'drome. Except for one gap (several hundred yards wide across the undershoot area of the longest runway) encirclement of the airfield was complete. He looked up with interest as a flight of three Hunter jet fighters screeched overhead in immaculate starboard echelon formation, before peeling-off in separate parabolic arcs. All terminated perfectly with a squeal of tyres and puffs of smoke, right on the button of the active runway. Swiftly decelerating as they disappeared along the runway, they turned off and headed to the squadron's respective dispersal points. Across the field Paddy could see the large hangars shimmering in the morning sunshine, their shapes distorted and hazy in the reflected heat. An engine coughed into life and the huge three bladed rotor of a Sycamore helicopter started to rotate slowly, blurring with motion into a huge shining disc.

Inside the two-room wooden shack a young airman by name of John Hollis, earphones clamped across his head, was listening to the two-way radio conversations between airborne pilots and Air Traffic Control. His job was to take loop bearings of such transmissions from aeroplanes and relay them to the Tower. These bearings would be used to advise pilots requesting Homing Assistance in adverse weather conditions, or in emergencies.

About to re-enter the small building, Paddy gave one more sweeping glance around the field and was somewhat surprised to see three civilian-clad youths sauntering unconcernedly and unchallenged through the large gap in the peripheral defenses. They were obviously making their way toward him. A minute or so later they paused before him, smiling and greeting him with the traditional,

"Yassou Gymvári!"

The time-honoured salutation and friendly grins subdued the corporal's doubts, so he willingly complied with their request for a drink of water. One after the other they swallowed the proffered drinks, courteously thanked the young Irishman and turned to leave. Like a perfect host with parting guests, Corporal Hale escorted them the few steps to the doorway. But he never lived to see the sunshine again. The last man of the trio to have thanked him for the refreshing drink whipped out a revolver and sent a bullet crashing into the surprised serviceman's forehead, killing him instantly.

The three gunmen then took to their heels and bolted for that opening at the runway end. Noise of the shot, having penetrated into the partitioned-off section at the rear of the two-roomed hut, prompted the radio operator to open a sliding hatch to investigate. His horror-stricken gaze froze at the sight of his comrade's body, slumped on the floor amid a widening pool of blood. Through the open door he could see three diminishing figures, running like startled hares for the gap in the wire fence. By the simple act of depressing a switch and speaking into his headset microphone, he was in immediate contact with the Duty Controller in the Tower.

"They've shot Paddy! They've shot Paddy!" he screamed to a startled listener, "Oh God! I think he's dead. He's bleeding awfully."

"Now wait a moment," said the Controller, "who shot who, and where? Calm down laddie, and give me all the gen."

Hysterically the radioman answered.

"Three men are running away from the Homer toward the fence. They shot Corporal Hale. All I can see is that one has a white shirt, one a green shirt, and the other is wearing a khaki shirt. Hurry, they're getting away!"

He was now crying openly, as he belatedly realised that he too would be dead like his companion, had the gunmen been aware another serviceman had been present in the rear of the radio shack. The concealing partition had undoubtedly saved his life.

It was at that precise moment in time I personally became involved in this incident. I was at the controls of a Sycamore helicopter, across the airfield from the Homer building. On the broadcast frequency I had just transmitted,

"Nicosia Tower, this is Heli Alpha requesting take-off clearance for a training trip to Tongou."

The reply I received electrified me.

"Heli Alpha, investigate reported shooting at the Homer building. You are cleared to cross the active runway."

My trainee co-pilot shot a startled glance at me as I practically stood the Sycamore on its nose in a full-powered transition from a three-foot hover. Seconds later we were hurtling across the wide airfield. Again my radio crackled to life.

"Heli Alpha, three men involved in shooting last seen running for the gap in the peri fence."

There followed a brief description of the shirt colours, and the fact that one of our own lads was victim of this atrocity. Members of the wire-stringing teams were now aware that something was amiss and had started to give chase. The fleeing fugitives were already adding to their impressive lead, and running like Olympic champions.

In less time than it takes to tell our steed was beyond the field boundary, skimming the undulating ground at close to 100 miles per hour. Scanning the horizon I spotted two figures briefly silhouetted to my right, before vanishing below crest of a small rounded hill. I skidded the protesting chopper around in a shuddering arc, now in hot pursuit. Moments later the same two figures popped into view again, their feet puffing up small explosions in dust of the arid plain. Aha, I thought, green shirt and khaki shirt, now we've got you! I closed in relentlessly. It was one thing however, to spot them…but how to affect a capture? Screeching as low as I dared, my tricycle undercarriage a foot or so from the sandy soil, I flung the helicopter at the fugitives, now glancing anxiously over their shoulders. At the last possible moment they threw themselves flat on the ground, my wheels brushing through empty space. As soon as I had passed they were up and running again on a different tangent, desperation adding wings to their heels.

A helicopter can be turned through 180 degrees in a remarkably short space of time and area but, in this case, I think we established a new world record. A climbing torque turn soon had our heading reversed, and down we swooped once more to harass the tiring runners. Again and again they were forced to prostrate themselves or risk serious injury as the menacing chopper swished overhead. Each lung-bursting sprint took its toll and eventually, completely spent, the exhausted duo went down to stay down. I brought the helicopter to a growling hover near the prostrate pair who now had their arms folded, hands clasped over the back of their heads to protect themselves from the blinding clouds of dust and grit being kicked up by whirling rotor blades.

Just a few minutes later some of the ground party arrived to complete the capture, so I set my machine gently onto the ground. The two young Greek men, in their late teens or early twenties, were no longer the debonair smiling characters who had so recently sipped cool water in Corporal Hale's presence. Chests heaving for breath, they stood dejectedly in their seat and dust-stained clothes, rivulets of perspiration making streaks down their dust-clogged features. Naturally curly locks were now as brown as the dun plain they had so fearfully embraced in their vain attempt to evade capture. A quick search revealed that their pockets were empty, no identification, no weapons, nothing. This alone spoke volumes, since everyone was supposed to carry some form of I.D. at all times. With their hands and feet securely tied, the luckless pair were bundled unceremoniously into the rear of our helicopter cabin and transported back to the helicopter pad closest to the Air Traffic Control Tower. A scant fifteen minutes after the alarm had been sounded, the first two captives were handed over to the waiting police contingent.

Two down and only one more to go, I thought. Now sadly aware that Corporal Hale was indeed no longer with us, but elated with our initial success, I took off again. Once more we headed through that tragic gap in the fence, determined to continue our search. The terrain was split by several dry river beds, shallow wadis, and rock-strewn slopes with sparse groups of trees, all baking in the merciless glare of the sun. Slowly we quartered the entire area, up and down, back and forth, methodically sweeping each square yard of the ground. We hovered and investigated each hollow, climbed again and moved on. In the mounting heat, magnified by the plastic cockpit canopy, my co-pilot Tony Harrison and I were soon drenched in perspiration. Our slow-crawling flight barely registered on the airspeed indicator and the cockpit temperature soared. An eye-watering hour ticked slowly by without success, no sign of our elusive quarry. The quest seemed hopeless, and our fuel-gulping helicopter would soon require its tanks replenished.

Suddenly, a flicker of movement in the topmost branches of a stunted tree caught my eye. Could it be? Yes by God! There was a figure crouching in one of the high forks! Down we plummeted like a stone, and I slammed the Sycamore onto the ground in a bone-jarring landing, whirling rotors mere yards from the gnarled tree trunk. Leaving Tony at the dual controls, I leapt out of the cockpit, un-holstering my sidearm as I went, and ran to the foot of the tree.

"Come down you bastard!" I yelled at the figure above me, "Or I'll shoot!"

My finger was white on the trigger, and my nerves more on edge than I'd ever imagined possible. It's doubtful the fellow perched up there could actually hear me above the roar of the helicopter engine, but my gestures with the cocked revolver were unmistakable. Chances were this final fugitive was armed, since a search of his two companions had not produced a weapon, but I felt in control of the situation and prepared for any eventuality. As he started to climb down, a bundled-up white shirt tumbled from the fork where he'd been crouching. It landed at my feet. Aha, I rejoiced inwardly, white shirt makes three down and none to go!

Once on the ground, the bare-chested Cypriot started to mumble that he didn't know what we wanted with him, that he was only bird-watching. It was not a very convincing act and a search of his pockets revealed nothing but a box of safety matches, British-made. Using a piece of nylon cord on hand for securing cargo, I trussed the youth's hands behind him and eased him into the rear cockpit, Tony covering me. I stayed in back with the prisoner while Tony performed at the controls to get us back to Nicosia Airport. A quick call on the radio informed all search parties that the three fugitives were all in the bag, so the search could be abandoned.

One hour and fifteen minutes after that startling distress call, the final prisoner was in the hands of the police. I arranged for the Sycamore to be re-fueled, and a few minutes later Tony and I were savouring the cold, delicious taste of a bottled Coke in the air-conditioned crew room. As we were preparing to remove our sweat-soaked flight suits, a burly individual wearing civilian clothes entered the crew room and identified himself as a Chief Inspector of the Criminal Investigation Branch,

Scotland Yard. On loan to the Cyprus police, he was now in charge of criminal cases such as we'd just been embroiled in.

Extending his congratulations to both of us on the success of our hectic flight he commented that, to his knowledge, this must be the first ever case of a one-hundred percent capture of three murderers by helicopter, and in record time too. He told us it would be possible, with a clever lawyer, for these men to walk free because they carried no identification, no weapons, and asked me if it would be possible to fly back over the same track covered before the first two men were captured, to try and recover the gun. Very few minutes later we were strapping ourselves into Heli Alpha and heading out toward the search area, this time at a more sedate pace.

The Chief Inspector sat in front beside me, and two of his assistants rode in the back. In my mind I tried to recall the rather hectic and jumbled impression of that first ground-scraping chase. About five minutes later, at the edge of a field with some stubble bordering the dusty open plain, I recognised the place our two sprinters had first thrown themselves flat as I buzzed them. Informing my passengers of this, I brought the Sycamore to a hover and slowly lowered the machine onto the ground in a helter-skelter of brown dust. It dissipated once I reduced power and the rotor blades slowed to a stop. The three officials disembarked, ready to carry out what I thought would be a hopeless task.

Suddenly one of the officers let out a whoop and pointed to a spot just about one foot from the wheel of the helicopter. There, projecting from dry soil, was the butt-end of a revolver! One of the gunmen, sure they were about to be captured, must have thrust the gun wrist-deep in the loose dirt. Down-wash from the helicopter's rotors during my slightly prolonged hover over the spot had no doubt uncovered it again. What fantastic luck! The Chief Inspector used a pencil through the trigger guard to lift what proved to be a Smith and Wesson 38 calibre revolver from the soil. He sniffed at it and said,

"Yes, it's been fired recently."

After another brief but fruitless search for more weapons we flew back to the helicopter pad, everyone feeling sure they now had a solid case against the three captured youths. Later ballistic tests proved this to be true. The revolver had been stolen from a serviceman during a previous incident, and had also been the weapon used in several other assassinations in Cyprus. In due time, and without me having to give evidence in court, the three prisoners were found guilty of murder and terrorism. They eventually paid the supreme penalty for their vicious crime. After the newspaper publicity I was reported to be on the E.O.K.A. hit list and posted away from Cyprus to carry on my flying duties elsewhere.

SQUADRON TRAINING OFFICER
Malaysia

Stationed in Malaysia, I was first attached to 194 Squadron based in Kuala Lumpur as the Squadron Training Officer. My job was to train pilots new to handling the Sycamore helicopters in the jungle theatre. Navigation over the dense jungle, and map-reading to avoid getting lost, were important skills to develop. So too was how to control the helicopter when descending into tight jungle clearings, and out of them, with little margin for error. Complete knowledge of one's aircraft and its components was essential. How far could one fly with a given amount of fuel? How much of a load could be carried with safety, and within load limits? How does one judge ground-speed, wind drift, and actual track by outside references?

Many of these lessons could be learned accumulatively by flying in the so-called Dry Season, when forward visibility was good, but during the Monsoon Season a better knowledge was required. Coping with poor visibility and with low cloud covering high ground, many geographic features which aided navigation would be obscured. All of these factors had to be taken into consideration, and took time for the pilots to absorb. In addition, pilots had to be trained to cope with emergencies like engine failure. There aren't many wide open spaces in true jungle territory, so pilots had to learn how to track along rivers where small islands might appear for an emergency landing.

Radio communication was also a problem in what was then called Malaya and only the jungle forts, manned by service personnel, had either a short runway where single-engine Pioneer aircraft could land, or at least a helicopter pad where a stable set-down could be made. These forts were also equipped with refueling facilities and had fuel supplies parachuted-in.

If a helicopter engine fails in flight there is only one sure consequence. The helicopter is going to start descending very quickly. It is essential that the pilot immediately lowers the collective lever if the engine quits, so maintaining a safe altitude above ground is advisory. If the collective is not lowered at once, the effect of drag immediately starts to reduce the main rotor's R.P.M. If allowed to continue like this, the main rotor becomes totally ineffective and the helicopter will accelerate more rapidly toward the ground, completely out of control. In powered flight, the main rotor accelerates air downward, through the main rotor. This can be noticed especially when the helicopter rises to a hover, after take-off.

In flight, if the engine fails and the pilot immediately lowers the collective lever, the pitch angle of each main rotor blade is rapidly reduced, and the rotor drag is decreased. Because the helicopter has now started a descent, this produces an upward flow of air through the main rotor system, and a free wheel unit in the engine assembly comes into play. The rotor R.P.M. will rapidly increase and can be used, when close to the ground, to greatly reduce the high rate of descent. If the landing area is clear and flat, a safe zero-speed can be achieved for touchdown. If the ground is not clear at the point of landing at least a zero rate of descent can be achieved and the chances of personal survival is present, though the helicopter would most likely break up as the rotors thrashed.

Part of my job was to ensure that every pilot was checked-out on fully-auto-rotational landings each month. This involved flying above the base airstrip at altitudes of 1,000 feet and actually switching the engine off, then an entry into auto-rotation, descent to just above ground-level over a selected, flat portion at the side of the active runway, and a nose up-flare to kill the forward speed. This, combined with a smooth upward pull on the collective lever, would arrest the rate of descent completely. Leveling the fuselage with a forward push on the Cyclic would result in a gentle, level landing, similar to a touchdown from a low hover.

It was essential that these emergency-landings were practiced regularly. As a result of demonstrating so many, and also sitting beside the other pilots doing the exercise beside me, I reckon I became quite expert at these emergency landings and could actually select a given spot where my wheels would touch the ground. At some public shows, back in Britain during open days to the public, I was often called upon to demonstrate this skill. One day at a Farnborough Air Show, I had to demonstrate this by actually landing on a flat-wheeled platform being towed by a low tug. I had to time my landing on that moving platform exactly when it was rolling in front of the judges at the show. I succeeded in my task to a generous round of applause from a huge audience.

'ANCOCK'S 'ALF-'OUR

At the Squadron Base in Kuala Lumpur one Sunday morning, I was in the office as standby-pilot, ready to undertake any mission that might be called. Looking out of the window I noticed a man dressed in civilian clothes walking around the Sycamore helicopter I had just prepared for flight. I walked outside, approached the stranger and, in a clear voice, advised him that he was trespassing in a restricted area. Kuala Lumpur was a jointly-shared Civil/Military airfield, and the Military section was clearly delineated. The well-dressed stranger came closer to me and in a quiet voice apologised.

"I suppose I should have come over to your office before looking at the helicopter."

As he was speaking he produced an identification document which revealed he was, in fact, an Air Vice Marshall of the Australian Air Force. His name was Hancock.

"I have just been posted in as the new Officer Commanding this Group."

We walked back to the office, where I had to be close to the telephone, and sat together to enjoy a nice cold Coca-Cola as he told me more about himself. Apparently he had never flown helicopters, although his experience was quite varied as to types of aircraft flown. I informed him that I was the Squadron Training Officer and sort of jokingly said,

"Well sir, if you want to convert to helicopters, which are the largest part of your new Command, I can convert you to type."

He was a very pleasant gentleman to chat with, very easy and relaxed. He did not have the loud, raspy voice I had usually associated with Australians. I liked him immediately, and thought I'd soon be seeing him again. That turned out to be a correct conjecture because the new boss was soon travelling around, getting to know his men and the general set-up of our command structure. It wasn't long before I was told that the Big Boss wanted to learn how to fly the Sycamore, and I was formally introduced by my Squadron Commander.

"Oh, we have already met," said the Air Vice Marshall, with a smile.

In the air, as my student, there was no sign of the difference in rank between us. He was a good student who quickly cottoned-on to everything I was pointing out to him. He helped me to relax and ignore the rank gap too, and we became friends. At this time, there was a rerun of an old British sitcom programme on television called, 'Ancock's 'Alf 'Our. It starred a comedian named, Tony Hancock…who always dropped his "Aitches" when speaking, otherwise the programme would have been called Hancock's Half-Hour.

One morning I was sitting in the half-crowded crew room waiting for my illustrious student to arrive, when one of the pilots who had just returned from a two-week leave asked me if he could take the Sycamore up for an hour of refresher flight, pointing to my machine.

"No," I said, "You can't 'ave it. It's reserved for 'Ancock's 'alf 'our!"

Little did I realise that, behind me, the men were now standing at attention and quieting. There was a little puff of warm air at my right ear, and an Australian-accented voice whispered,

"'Ancock…is 'ere!"

There was a loud roar of laughter, enjoyed also by the new C.O. He too had a good sense of humour, it turned out. The training trip went well and I sent the Air Marshall solo that day. I flew him on numerous occasions when he had to go somewhere in a hurry, and we always put the dual controls in so that he could keep his hand in. When he was posted back to Australia, everyone commented on the fact that he had been a good boss, well-liked by all.

A number of years later I was serving my third tour of duty in Malaysia, my seventh year in that theatre, and based at Butterworth near Penang, in northern Malaysia. I was now Training Officer for 110 Squadron. One day, I was informed that I would have a V.I.P. trip the following morning. I prepared for it as usual and, with my brass polished, waited beside the Group Captain commanding the base. He was a burly Australian, with a sort of pugnacious attitude. He told me that the V.I.P. was his Big Boss and he was looking forward to bending his ear during the flight to the military hospital at Taiping. Now I knew where I was going…an easy forty-five minute flight, along well-recognized routes.

When the four-engine transport drew up and shut down its engines, doors opened, stairs were put in place and the V.I.P. descended. He was dressed in the uniform of the Marshall of the Royal Australian Air Force, but he was none other than my former pupil! When it came to introductions, Mister Hancock stepped out of line and came up to me with his hand outstretched.

"Mac! What are you doing here?"

He was as surprised to see me as I was to see him, in his reflected glory as the Number One Man in the

R.A.A.F.! I jokingly remarked that we did not have the dual controls in the Sycamore we were to fly in and his attitude sort of confirmed that he liked the idea.

"Ten minutes only sir, and we will be ready."

Group Captain Case was not amused at losing his chance to bend Hancock's ear, but what could he say? That evening, my wife and I were delighted to be seated at the V.I.P. table, and enjoyed a fabulous meal.

CASUALTY IN THE JUNGLE

During WWII, the Japanese army successfully invaded many countries in the Far East including mainland Malaya, now Malaysia. A number of patriotic Malayan citizens quickly formed their own Anti-Japanese Army and disappeared into the leafy but hostile shelter of the jungle. There they would wage their own version of unceasing guerilla warfare against the Nipponese invaders. They used the initials M.P.A.J.A. Many in this underground organization were of Chinese origin and already steeped in the doctrines of Communism. The most hard-core used every opportunity available to convert their comrades to their political beliefs, then demanded complete and utter loyalty to "The Party."

When the Japanese finally surrendered in 1945 and a shaky peace was restored, the combined political parties in Malaya set to work, under British rule, to rebuild the sadly damaged structure of their country. Their hopes of returning to the harmony of pre-war days were rudely shattered when fanatical members of the disbanded Anti-Japanese Army came out openly in favour of the Communist Party, and voted so. Shortly thereafter that doctrine was outlawed by Parliament, forcing some of the disgruntled "Reds" to go underground again, using their recently-acquired expertise in a bitter campaign against their former tutors and allies. Anyone who publicly refuted Communism became an enemy, and a possible target for assassination. The worst period of this political unrest was undoubtedly the years between 1946 and 1959, when most casualties were inflicted upon either side of this bitter struggle. Those hard years took a heavy toll of life amongst military and civilians alike. Apart from the toll of wounds received in battle the jungle itself (and its strange inhabitants) added their own bizarre quota of casualties to the long list. Private Frank Burdett, a young soldier from New Zealand, was one such casualty.

JUNGLE PATROL

Early February 1959, a platoon of New Zealand soldiers serving with the combined British/Malay forces were airlifted by helicopter to a small aboriginal settlement deep in the central highlands of Malaya. Their task was to seek out and capture, or kill, a band of C.T.'s (Communist Terrorists) known to be operating in that remote area. They were also to attempt severing an enemy courier line running North-South between cells in the rebel organization. Watched by the friendly Temiar tribespeople (to whom the helicopters were now a familiar sight) the soldiers assembled, strapped on their heavy backpacks and slowly filed into the jungle, following a trail that wound steadily upwards into the Cameron Highlands.

Ever alert for booby-traps or signs of human passage, the "Kiwis" finally penetrated into suspect territory. In the eerie semi-twilight of jungle floor they were soon saturated to the skin with sweat. Their green uniforms already showed mute evidence of the unfriendly embrace of creepers and vines, all armed with formidable hooks and thorns fiendishly designed to rip clothing and skin indiscriminately.

"Who says the jungle is neutral?" muttered one discomfited trooper, as he slipped and slithered on one particularly steep incline.

In this strange environment one's estimation of distance travelled could be very misleading. Each forward step has to be won the hard way and progress is often measured in yards or feet per hour. Late in the afternoon of their second day's march, the platoon leader was relieved when they finally stumbled into a small clearing created by a rock slide from a limestone crag jutting above them. This might give him the opportunity to ascertain their present position with reasonable accuracy. After a wary search, blinking in the bright sunlight, the men shucked-off their backpacks and turned to check each other out for leeches or cuts, which could turn septic in a remarkably short space of time. First-Aid and personal hygiene attended to, they enjoyed a brief snack and rested their weary muscles. Climbing up to an exposed ridge on the shoulder of the crag their leader was able to take several compass bearings on some distant but prominent peaks, thus fixing their position with greater accuracy.

Setting up his transceiver, private Burdett established contact with base camp and relayed an updated progress report, including the map grid references of their present position. Ending his transmission with a cheerful,

"All's well here. Will transmit again at 08.00 hours, tomorrow," Burdett signed off and repacked his equipment. Little did he know then how much that next transmission would mean to him, and that he would not be operating the set for the morning broadcast. The next time he would enter that small clearing would be in a dramatic fight against time in an attempt to save his life.

"Come on lads, we can still cover some more territory before we make camp for the night," said the group leader.

Once more the troops filed into leafy twilight. Night falls with amazing abruptness in the tropical jungle, but the seasoned troopers were well prepared before darkness enshrouded them. A fire was out of the question, since it could give away their presence to a nearby enemy, so a cold meal was consumed. Flimsy shelters were built and sentries were posted around the perimeter of their small campground. It wasn't long before bone-weary soldiers turned in for some well-earned rest. Except for the usual night noises of the jungle all was quiet, and the small encampment slumbered.

JUNGLE MARAUDER

Alerted by an unerring sense of smell, a quiet marauder crept through the darkness, heading directly for the unsuspecting soldiers. Closer and closer this ghostly figure slunk, until penetrating the camp's perimeter, moving stealthily toward slumbering troops as silent as death itself.

In his little one-man basha (a shelter made of ferns and small branches) private Burdett stirred uneasily, drowsily becoming aware of a presence beside his bed, shaking the flimsy structure.

"What the blazes do you want?" he sleepily inquired, thinking it was probably his turn for guard duty.

The answer was a deep electrifying growl that set his hair on end. Fully awake now and aware of a pungent animal odor, he stretched out his arm to reach for his rifle but screamed out in agony as his shoulder was seized in a bone-crushing grip. Moments later, the luckless radio operator felt himself being dragged bodily through the collapsing walls of his shelter. He screamed again, in mortal fear, until a blow from a massive paw silenced his hysteria.

For a few moments panic and uncertainty reigned in the camp. The rest of the soldiers, fearing an enemy attack was in progress, were reaching for their weapons in the stygian darkness, and tumbling to the ground as they gathered their wits about them. Suddenly one of the Kiwis in a moment of inspiration turned on a powerful flashlight, sweeping its beam around in a wide arc. This revealed a scene that chilled their very hearts. Transfixed in the bright light was an immense tiger, tail twitching fitfully, its eyes balefully aglitter, reflecting two enormous emeralds. Dangling from its powerful jaws like a rag doll, hung the blood-soaked body of one of their comrades. An ominous growl emanated from the huge beast's throat as it slowly backed away, still retaining its prey. For a few brief, measureless blinks of time no one moved or spoke. It was impossible to shoot without endangering each other or the life of private Burdett, if indeed he was still alive.

One guard suddenly loosed-off a rapid burst of automatic fire directly over his own head into the tree canopy. This had the desired effect. Startled by the gun flashes and the sudden noise, the striped monster dropped its limp victim and with two enormous bounds disappeared into darkness. Carefully the unconscious soldier was laid onto an unrolled poncho and examined in the light of several flashlights. Immediately apparent to all were the deep gashes across his face and abdomen, as well as the badly chewed-up shoulder, still bleeding copiously. Curtly the troop commander issued orders as he applied field dressings to the worst wounds, and applied First-Aid to the best of his limited ability.

"I'll need some of you to backtrack down the trail to the clearing we rested in today and start felling some trees. We have to make that clearing big enough for an R.A.F. helicopter to land in and prepare a pad for it to set down on. Also have some smoke signals to assist the pilot in finding us."

Quickly he nominated those who were to go and reminded them to blaze the trail for those who were to follow up carrying their injured buddy.

"Cut branches and lash several ponchos together to make a stretcher. We'll all take turns at carrying Burdett back to the clearing."

Turning to a corporal he knew was qualified to operate the radios, he quoted the message that was to be relayed at 08.00 hours. This included map reference of the proposed helicopter landing site. Silently he gave thanks for the undoubted accuracy of that location. The next day I would be similarly grateful. In the early hours of a Sunday morning the troops prayed briefly for their injured companion, then set off for the grueling march down the tortuous trail.

AN INTERRUPTED FAMILY OUTING

That same Sunday morning, after breakfast at home in Petaling Jaya (close to Kuala Lumpur) my wife Bunnie and I decided to take our children for a swim at the R.A.F. base. It was meant to be a lazy day's outing, with a picnic lunch thrown in. This decision unwittingly involved me in private Burdett's struggle for life. Having arrived at the pool and secured a snug picnic area in the shade, I told my wife I'd just pop around to 194 Squadron to see if any mail from home had come in over the weekend.

"Don't be long," Bunnie said, as she busily undressed the baby.

"Won't be a jiffy," I replied, knowing the offices were just a few minutes' stroll from where we sat.

As I checked the mail slots in the pilots' crew room, a young corporal ground crew man on duty informed me that the two pilots on standby duties had been called out earlier on emergency flights to the south of K.L., and would be away for most of the day. No sooner were the words out of his mouth when the squawk box connected to the Operations Room erupted into life.

"Are there any other serviceable helicopters available at 194 Squadron?" queried the Duty Op's officer.

The corporal nodded yes to me, holding up a notice board that revealed all I needed to know. Depressing the talk button on the intercom, I replied,

"Affirmative, Sycamore XG529 is serviceable, and has 65 gallons of fuel in the main tank."

Recognizing my voice as that of Squadron Training Officer, the Duty Controller muttered a thanks-be to God and rapidly filled me in on the casualty evacuation flight, so urgently needed.

"The helicopter at Ipoh, which would normally have done this mission, has developed an engine problem and won't be available for hours. Can you nominate a pilot we can call in to do this?"

Without hesitation I replied, "Yes, I'll do this mission myself."

"Good," said the Op's Officer, "Dr. Ross will be on his way down in a minute."

He passed me all the important map references I'd have to locate and land in, and with a heartfelt, "Good luck," he signed off.

Scanning the large map of Malaya pasted on the crew room wall, I quickly pinpointed the location of the new LZ (Landing Zone) and transferred it to a spare map, which I stuffed into my pocket. Since my flight suit was home being laundered I chose to fly dressed as I was, in brief shorts and a striped blue and white tee-shirt. On my feet were open-toed sandals, totally unsuited for jungle trekking should I have to resort to that. I shrugged, grabbed my helmet from my locker, and raced out to the machine awaiting me on the dispersal pad. During a rapid but thorough pre-flight of the Sycamore I confirmed the main tank was full to the brim, holding 65 gallons of fuel, and the reserve tank, empty. Two minutes later the huge rotors were turning and the medical officer was ducking his way into the cockpit to sit beside me. He was, in fact, still fastening his seat belt as wheels left the ground in a vertical swirl of dust.

It was a beautiful morning for flying, clear blue skies, small puffs of white cumulus cloud casting round shadows on the ground, light winds, and excellent visibility.   Quickly our helicopter sped across the sprawling outskirts of K.L. as Flight Lieutenant Iain Ross finished strapping himself into the passenger seat. Mountains to the northeast seemed to have been etched against the horizon like a huge painted canvas. Using the airborne radio frequency, I informed Operational Control that, to save valuable time, I intended to fly a direct straight line course to my objective.   This was a significant departure from the usual technique of following well-established routes such as rivers, roads, railway lines, or clearly delineated valleys, where map-reading was much easier and there were numerous landing spots in cases of emergency or inclement weather, should the need arise.

Having obtained their official blessing, I lofted the eager Sycamore to an altitude of 5,000 feet and bee-lined toward the Cameron Highlands. Away to the west I could see mangrove swamps marking the shoreline, and the reflected glitter of sunshine on the sea in the Straits of Malacca.   Soon the suburbs of the capital city disappeared behind us and were replaced by the rising ground of the central spine of Malaya's mountainous ridges. I made my first of many position reports when overhead the resort of Fraser's Hill. This was composed of several buildings perched on a high plateau, marking the point where a tortuously winding road ceased climbing from the west and commenced a similarly twisting descent to the east, into a central bowl where tiny kampongs (villages) nestled among the vivid green patches of paddy fields. Natives stooped to their labour, knee-deep in muddy water.

My course now took us over more virgin jungle, where tenacious creepers and assorted vines festooned all but the sheerest limestone crags, jutting out like fangs underneath our speeding wheels. Viewed from the air at close quarters, the jungle canopy presents an almost unbroken vista of greyish greenery, not unlike thousands of immense cauliflower or cabbage heads carelessly tossed together. Occasional reflected sunlight glinted off the surface of rivers winding their lazy way to the sea. To me, each river and mountain peak represented familiar landmarks where I could update our rate of progress across the map and report my position to the listeners at Op's Centre. I was constantly doing mental arithmetic, computing the distance travelled, fuel consumed, and the distance still to cover, as well as the all-important timing. Almost two hours after departing K.L. we swept over the huge tea plantations of the Boh Valley, almost 7,000 feet above sea level in the Cameron Highlands. Here during happier times Sultans, Princes, and tourists came to play golf, and enjoyed burning fires in their snug holiday cabins to ward off the chilly night air.

DECISIONS IN FLIGHT

The Sycamore helicopter engine consumes about 25 gallons per hour in cruising flight. Calculating that, since take off we had used up about fifty of our original sixty-five gallons, leaving fifteen gallons, which was enough to keep us airborne for another thirty-six minutes. Glancing at my map I decided to divert to Fort Brooke, not very far from our direct line of flight, to replenish my dwindling fuel supply. Notifying Operations about this decision, I started a slow descent into a natural amphitheatre created by the conjoining of three separate valleys with steep heavily-wooded slopes on all sides. Fort Brooke, named after a young police lieutenant killed by the terrorists, sits on a rounded knoll around which loops a fast-flowing river. It is one of the few jungle outposts that didn't have an airstrip because of its natural topography, but sported two helicopter landing pads. As we alit on the top pad, Inspector Jock Currie (in charge of the post) came ducking under the whirling rotor blades and greeted me with his usual cheerful grin. Quickly I filled him in on the nature of our mission and requested an extra-fast refueling. His face fell as he related news that had me going over my calculations again with renewed urgency. Apparently a recent air-drop of aviation fuel, drums attached to parachutes, had gone awry. Some chutes had candled (not opened) and the resulting heavy ground impact had ruptured the drums. A strong cross wind had also contributed to this loss, blowing some of the chutes across the river. The parachute shrouds had tangled in the high trees. Ruefully Currie admitted,

"I haven't a drop of aviation fuel to give you."

Fort Brooke

Perusing my map again with extra care, I again estimated distance, time, and fuel requirements to reach my destination. If the grid reference of the casualty clearing was accurate (all too often they were not) if I could locate it without delay and, if it was of sufficient size to permit safe entry and exit (very often they weren't) I estimated that I could just do the pick-up and then fly onward to Fort Chabai, some thirty miles distant to the north, that is, if we didn't meet a head wind and if the fuel gauge was reading accurately. That was a lot of ifs to stake human lives on, including my own! Telling Jock to inform base of my intentions, I swiftly restored the rotor RPM from idling to operational and without further delay, lifted off.

Using minimum power where possible, I snuck the helicopter over several rises to finally enter the valley of the Sungei Perolak, zooming over the bamboo longhouse where the Kiwi troops had assembled just three short days ago. In mere minutes we covered tedious miles the troops had so patiently trodden. A welcome sight greeted my anxious gaze; a plume of white smoke was rising vertically from the trees ahead. Spiraling almost straight upwards, it indicated an almost non-existent wind condition. I knew from experience that this would allow me to choose the easiest approach for a descent into the new clearing, which proved to be small, uncomfortably so. My lightweight condition however, allowed me to perform a controlled slow vertical descent, alighting gently on the jungle floor. Iain Ross was out of the cabin almost before the landing was completed, and made his way to the waiting soldiers. Moments later he returned and informed me that the troops carrying the injured man had not yet reached the clearing. He was still being carried down the trail, and wasn't expected for at least another twenty minutes or so.

Here was yet another dilemma for me. To leave the engine idling for that length of time would use up too much precious fuel, and certainly we wouldn't be able to reach Fort Chabai. An F.E.A.F. (Far East Air Force) standing order also forbade the shutting down of helicopter engines in suspect territory, especially in clearings not large enough to accept more than one chopper. Another consideration was the fact that the Alvis Leonides engine fitted in the Sycamore was notoriously temperamental and difficult to restart after a brief shutdown in warm climates. This was just another spot decision to be made. The New Zealanders all looked up in surprise as the engine growled into silence and the huge wooden rotor blades drooped to a standstill. Their eyes opened wider when I stepped out of the cockpit, dressed in my natty sports attire! Doctor Ross was escorted out of the clearing to head towards the patient.

From start up in Kuala Lumpur to shut down in the clearing, the engine had been running for two hours and twelve minutes. Calculating at 25 gallons per hour, we should have consumed approximately 55 gallons, leaving me with an estimated 10 gallons. Fort Chabai still lay thirty miles to the north. At 85mph cruising speed, it should take us 21 minutes to cover that distance. The ten gallons of fuel remaining (if the gauge was correct) should keep us in the air for an estimated 24 minutes.

"You're cutting things a bit fine, Jim boy!" I said, and forced myself to relax.

Eventually the exhausted stretcher crew stumbled into sight with Dr. Ross already administering to the young soldier. Once the patient had been transferred to the rear cabin, and plasma bottle etc. strung up, Doc Ross rotated his seat to face rearward and continued to monitor Burdett's vital signs.

"He's really in a bad way and must get proper treatment as soon as possible," he said to me.

Gravely I acknowledged this and started my drill for boosting the engine. Mentally I crossed my fingers and, with a muted prayer, switched on. It was sweet music to my ears when that powerful engine burst into lusty life at the first attempt! Moments later we were rising out of the trees and on our way north. To proceed in a straight line would have meant climbing over higher ground, already being capped with domes of towering cumulus clouds, guaranteed to create extreme turbulence that would distress my passengers and increase the fuel consumption. I chose to head slightly east down the valley, where we would encounter lower contours before continuing on the northerly heading. My eyes kept flickering from the trees ahead to the fuel gauge, now reading dangerously low at just above the red line marking zero.

The endless carpet of trees seemed to trundle past at a snail's pace, while the minute hand of my wrist watch seemed to whiz around like it was the second's hand. Time flew while we crawled across the verdant landscape. Butterflies started to do a fandango in my stomach as the fuel gauge sank ever lower. Suppose my calculations were wrong! I thrust the unwelcome thought from my head and concentrated on flying as smoothly as humanly possible. As low as we were, just above the treetops and on the lee side of the mountains, we were out of radio contact with the base. Transmissions would go unheard except perhaps, by some high-flying passenger liner passing overhead. Like a particle whirling in its own galaxy we were alone, dependent on fate and my own decisions. Occasionally the helicopter would overfly tiny open gravel bars in the loops of a winding river, and the temptation to set down was almost overpowering.

"The soldier will surely die if we delay getting him to hospital for treatment," I found myself muttering. An inner voice would then counter with, "We'll all die if you don't set this helicopter down before the engine runs out of fuel."

Tension in the cockpit mounted. Even Dr. Ross fell silent as he stared around at the passing panorama. I found myself leaning forward, as if to urge greater speed.

Some eighteen long minutes after popping out of that little clearing a welcome sight came into view in the form of a huge red scar, standing out clearly from a green slope directly ahead. This was a prominent landmark used by all aircrew flying into Chabai. It had been formed by a long-ago landslide. Fixed wing pilots using the short take-off and landing Pioneer aircraft habitually turned onto a westerly heading over this gash in the trees, and let down into a narrow valley for a short distance before committing themselves to a ninety-degree right turn into another cleft from which there was no turning back. The airstrip at Fort Chabai would then be in view, in a narrow cul-de-sac that ended in a steep incline. There was only one way in. The westerly turn into the valley was a great lift to my spirits. My navigation had been perfect, but the fuel gauge was now reading zero and the needle seemed to be hanging against the stops, as if to warn me of our predicament. For a heart-stopping moment the engine spluttered...and then resumed its even-tempered purr, as my knuckles whitened on the controls. Below the helicopter now, a raging white torrent foamed madly between huge rounded boulders. As I watched, an immense tree trunk tossed like a cork in the surging water and posed prettily in the middle of a rainbow spray before vanishing from sight. Hardly the best place to suffer an engine failure.

The right-angled turn and the welcome sight of the grassy runway at Chabai slid into view. Avoiding any abrupt movement of the controls or changes in attitude which could cause any of the remaining fuel to surge in the main tank, away from the immersed fuel pump, I gently wheeled the Sycamore onto the ground in a running landing without attempting to hover. We taxied clear of the active strip into a holding pen, and left the runway clear for any other arrivals. A numbing cramp in the fingers of my right hand made me aware for the first time just how tightly I had been grasping the cyclic control during those last few minutes of flight.

ANTI-CLIMAX AND HAPPY ENDING

Switching off all systems and waiting till the rotor blades ceased spinning, I turned to Iain Ross and gave him a big relieved smile. The Chinese police inspector in charge of the fort then approached and was duly informed of our needs. I asked for the main tank to be refilled to its capacity of 65 gallons, and requested an accurate count of just how much fuel his men would require to do this. As the staff fulfilled this task Dr. Ross and I stretched our legs, walking around a little to unwind. We were treated to a much-needed drink of fresh cold mountain water, which tasted like nectar. The patient was carefully monitored and seemed to be holding up remarkably well, considering his plight. Final calculations when the tank finally brimmed full, revealed that we had used-up 64 of our 65 gallons of fuel. The one gallon remaining would have been enough to keep us in the air for approximately two more minutes, or to cover three more miles of territory before fuel starvation. That was cutting things much too fine for comfort.

The remainder of the flight to Tai Ping Military Hospital was completed in record time by again making a straight bee-line course, with good radio contact. The hospital staff awaited our arrival at the helipad, and the patient was quickly wheeled into an operating room without delay. Doctor Ross was still in attendance. My helicopter was refueled and I departed along regular routes for the return flight to Kuala Lumpur, leaving private Burdett in capable hands. Some six and a half hours after I left my wife and family, I rejoined them at the swimming pool to be greeted by my very irate wife.

"And just where the hell have you been all this time, eh?"

She forgave me when I told her the story of private Burdett's ordeal. I'm pleased to tell you, although he was badly scarred, Frank Burdett survived his ordeal, and was repatriated to his home in New Zealand.

****

Many years after this tiger casualty evacuation, I was emailing with another Kiwi soldier I'd met on flight duty. He asked me what areas of mainland Malaya I had been working in. I replied that my flying duties took me all over that peninsula, but the many operations had been in the north, close to the Thai border. I told him that I brought out another Kiwi soldier who had been viciously mauled by a man-eating tiger. We found out this same tiger had killed five aborigines of the Temiar tribe. Rather excitedly he exclaimed,

"I was talking to him a couple of days ago on email! His name is Frank Burdett!"

I asked my new friend if he would contact private Burdett and give him my email address, explaining that I was the pilot who had flown him out, that particular Sunday morning. Eventually Frank and I established contact and, after all these years since the event, we now exchange daily messages.

A MODERN RESURRECTION

Early 1960's

The locale of this true story is the mainland of Malaya, one of the most enchanting countries in Southeast Asia. Malaya stretches out like an inverted thumb from its border with Thailand in the north, to the island of Singapore in the south. Clad for the greater part in a thick green mantle of tropical jungle and mountainous rain forest, it is a land of great beauty and vivid contrast. Rich in massive deposits of almost-pure tin ore, Malaya is also liberally endowed with huge deposits of bauxite, gold, and silver. Large tracts of land have been converted into many neat plantations for the cultivation of coconut palms, oil palms, tea, and rubber. All of these industries, plus vast sources of valuable timber such as Teak, help to give the inhabitants of this country a high standard of living when compared to their less-fortunate neighbours.

The wet tropical climate also encourages the growth of assorted fruits such as bananas, pineapples, papayas, mangosteens, rambutans, and the supposedly aphrodisiac durians, much favoured by the Chinese population. Using modern techniques and cross-pollination methods, numerous varieties of exotic orchids found growing wild in the jungle are also cultivated in nurseries, forming yet another lucrative business. Thanks to the benevolent sequence of tropical sunshine and torrential monsoon rains, rice (staple of most Asian people) can be planted and harvested twice within each calendar year. With the exception of a central plain encircling the capital city of Kuala Lumpur, most of the inhabitants reside along the eastern and western coastal fringes, separated by two lengthy ridges of central mountains, partitioning the northern two-thirds of the peninsula.

Each year advances of civilization encroach upon the verdant jungle, forcing it to retreat inwards as if in flight to the lofty hinterland. These dense jungle areas were for years, a battle area for opposing forces. When the Japanese invaded Malaya in 1942, a number of British soldiers and patriotic civilians escaped into the jungle, waging guerrilla warfare against the hated intruders. Calling themselves the Malay Peoples' Anti-Japanese Army (M.P.A.J.A.) they became wise in the art of jungle warfare and survival. They were a thorn in the side of that enemy until the war's end, in 1945.

Unfortunately, the cessation of hostilities did not restore a lasting peace to Malaya. Many members of the M.P.A.J.A. came out openly in favour of Communist doctrines and were subsequently outlawed for those beliefs. Disillusioned, they returned to the now familiar shelter of the jungle to vent their wrath and bitterness against the forces of the joint Malay/ British government. During this troubled post-war period, helicopters of the Royal Air Force were called upon to act as aerial taxis for the Crown, giving the troops a fast, flexible radius of action, vastly superior to that of the jungle-impeded rebels. As successive sections of the country were cleared of communist terrorists (CT's) the harassed survivors retreated into a pocket of jungle just south of the Thai border. Little was known about this seldom-penetrated area, but it soon became the focal point of large-scale operations designed to end the long insurgency.

A platoon of Allied soldiers operating in this section of the Ulu became aware that they were being stalked as they moved around in green twilight of the jungle. Occasionally there might be tantalizing glimpses of ghostly figures which seemed to dematerialize and vanish, wraith-like, into the undergrowth when approached. Now and again the soldiers would stumble upon crude lean-to shelters bearing mute evidence of recent hasty evacuation. It was decided that these phantoms must be a group of aborigines previously unknown to exist in this area. They were of very small stature, not warlike, or unfriendly. No signs of booby-traps or terrorists so far, so the soldiers started to leave

175

parcels of food and small gifts for the nomads, and were rewarded one day when a group of the tiny people emerged to greet them.

Much smaller in stature, and decidedly swarthier in colour than the Temiar natives to the south, the pygmies gathered around staring in awe at the tall, green-clad warriors. Wonderingly, they fingered the fabric of jungle-green uniforms and examined the weapons and trinkets adorning the men. They themselves were clad only in a brief loin-cloth made of some bark-like material. They carried only small wooden clubs. No other weapons or blow pipes (such as were used by other tribes of aborigines further south) were in evidence so there was no danger of poison darts being used here. Now that friendly relations had been established, these tiny native people soon became an integral part of camp life, and even assisted as scouts when the troops were on patrol. Initially, whenever a helicopter clattered noisily onto the landing pad they were terrified, but soon became accustomed to such events, even assisting in loading and unloading supplies. Nimble fingers were soon put to use in weaving huge mats of bamboo to cover the helicopter landing pad, greatly adding to the safety and durability of same. They also acted as sentries around the perimeter of the camp, guarding against intruders.

One of the regular helicopter pilots serving the area, I alit gently onto the mat one day and shut down the engine. As the rotor blades slowed to a standstill I became aware of an air of gloom pervading the camp. Gone were the usual cheerful smiles. When I enquired, "What's the matter?" I was informed that the small natives were planning to vacate the area. The reason for this apparently, was that one of their favoured youths had come down with a mysterious case of poisoning which threatened his life. It was believed that, if he actually died there, his ghost or spirit (Hantu) would forever be imprisoned there, so they felt it necessary to move on. Naturally we could not just allow this to happen without doing everything within our power to save this likeable youngster.

Because of my ability to fly like the giant hornbills of their forest I was held in high esteem by the tribespeople, so I was chosen to be spokesman for this occasion.

Considering we had no common language, this was not an easy interview. Using much waving of arms, pointing to myself and the boy then to the helicopter, we finally conveyed the message that I wanted to take this sick boy with me in the helicopter to see if we could save his life. After deliberation, the senior men of the tribe agreed to this airlift and the semi-comatose boy was lifted on a stretcher and brought to my Sycamore chopper. A ceremony then began which moved me deeply. Each member of the tribe, young and old, filed past the stretcher and touched the youth lightly on the forehead. It was a moving gesture which seemed to combine all the elements of tenderness and get well wishes with a final farewell salute. Plaintive notes of a bamboo flute quivered in the air around us, somehow reminiscent of a lonely kilted Highlander playing a bagpipe lament on a heathered Scottish hillside, as the stretcher was lashed to rings on the floor of the helicopter's cargo/passenger area.

All secured and the doors closed, I roared the engine into life and lifted off, spiraling vertically upwards for about 150 feet before heading on my way to the hospital in Ipoh, close to the military brigade headquarters. Winging our way southwest at one hundred miles per hour over the lumpy grey-green canopy of the jungle, I kept track by referring to my map and picking out various landmarks such as river junctions and isolated peaks, all as familiar to me now as the street of my hometown. The shimmering mountains on the horizon swam into focus, each one capped with fluffy white heaps of cumulus cloud. Rising to a little over 5,000 feet above sea level, they marked the last physical barrier between our present position and the tin-rich Kinta Valley, wherein lay my destination. Some thirty minutes later, I radioed my position as being overhead the short airstrip at a jungle fort named Fort Kemar, and gave my estimated time of arrival for Ipoh airport. I relayed the pertinent information, that I would require an ambulance to meet my helicopter upon landing. Details about the aborigine boy were included in my report. I skimmed at low level over the trees, keeping clear of clouds, as I navigated my way into a narrow pass between towering peaks. This would lead me directly toward the Ipoh plain and my destination.

On the downslope I passed a huge Flame of the Forest tree, looking as if it were indeed on fire. This was a familiar old landmark which assured me I was on the right path, in the correct valley. I mused how strange it was, how a single tree could aid navigation over the jungle. I had used it often enough in that role, and taught it to the new pilots I trained. Ahead of me I could see huge scars of tin mines marring the level plain. Here

179

Man's search for mineral wealth had simply gone mad. Mammoth dredges, looking like the old-fashioned Mississippi steamboats with their stern wheels churning, were chewing their way across the land. They gouged out huge bites of tin ore, leaving behind quarries filled with deeply-coloured pools of water. From my vantage point they looked like a horde of prehistoric monsters, intent upon destroying this entire land.

The main highway running north and south wended its way between huge outcrops of limestone. They were each honeycombed with caves, many of which housed brightly-decorated temples, built directly into the living rock. Presently, the sprawling outskirts of Ipoh slid beneath our speeding wheels and I set up my final approach to the landing pad where I could see an ambulance, already waiting. The sick youngster had shown little interest in his surroundings as we sped over the high jungle, but became agitated as we neared this town displaying strange sights he could not decipher. His little face was grey with fear after landing at the airport, and he became almost unmanageable when it was time to transfer him to the ambulance. He clung desperately to my hand, and with eyes widely appealing, strove to maintain contact with the one face he knew...mine! There was only one way to pacify him, and that was for me to accompany him to the hospital. I secured my machine and climbed into the ambulance.

After he was mildly sedated, I left him to the tender care of the doctors and nurses, returning to the airfield. I refueled and left for my journey to Squadron Headquarters, Kuala Lumpur. Later enquiries uncovered that the youngster had an infection from some mysterious bite, possibly from a snake, but was recovering well with the aid of antibiotics and modern hygiene. Several days later he was fully restored to impudent health and walking around the wards dressed in khaki shorts, a shirt, and flip-flop sandals on his feet. He soon became the pet of staff and patients alike. Sometimes he was taken into town, where he saw high-rise apartments, buses, taxis, and market stalls, where food of all descriptions was on display. His palate, used to a simple diet of roots, fish, grubs, and occasional pieces of raw meat, soon grew accustomed to the staple foods offered at the hospital. He especially liked ice cream and chocolate, quaffing it as often as he could beg, borrow, or pilfer. He even attended some evening film shows at the hospital, where the language could vary between English, Tamil, Malayu, or Chinese. It didn't matter to him. He would just sit and stare, round-eyed, like any child, although he could not make head nor tail of whatever was being said.

Eventually, a couple of weeks after his traumatic arrival at the hospital, it was decided that he was fit to go home. But how would that be accomplished? He had no authority, or priority, to be taken in a helicopter all the way back toward the Thai border. The centre of operations had now moved on from the area he had come from. In desperation, one of the head principals at the hospital made a call to 194 Squadron offices and requested to speak to me. Once aware of the situation, I informed him that I was the Squadron Training Officer and had a new pilot to train in jungle procedures, as each new arrival had to undergo fifty hours of flying before qualifying for general operations. I told him that I would make a point to fly back to Ipoh, and would notify his office as to date and time of our arrival. Part of the new pilot's training would be to navigate his way to the area where the lad came from, and we would deliver the boy back to his tribe. No other authority was required.

Flying a turbine-powered Sikorsky S.55 helicopter, we flew out of Kuala Lumpur and headed north towards Ipoh. The new pilot was working on his map-reading and navigation etc. I pointed out various landmarks and routes we followed, with explanations as to various procedures. Arriving at Ipoh around lunch time, we stopped over at No. 2 FIB Headquarters, ate, and informed the Staff Op's of my intended journey. We also volunteered to take any supplies or mail that had little or no priority with us, delivering it wherever required. An offer like this didn't come along often, so the Staff was only too happy to accept...never look a gift horse in the mouth, as they say! So with one army passenger, bags of fresh mail and supplies, we headed to the hospital to pick up the young boy. He was delighted to see me, and to know that he was now going to be flown back to his tribe. I briefed the army passenger on our intentions and instructed him that, if we did land at the same clearing, he was to keep the doors closed and the youngster quiet until I personally opened the door. I told him we didn't know if the tribe would still be there or if we would have to search around.

We took off and headed out over the Kinta Valley, where I pointed out the details to the new pilot. Reversing our track into the valley, climbing toward the gap that led to Fort Kemar, I showed the newcomer what a Flame of the Forest tree looked like, pointing out its usefulness in navigation. Doing well, my co-pilot followed the line marked on his map, navigating successfully all the way until we arrived in the former operational area. In due time we flew overhead the old clearing, the matting on the landing pad showing up well. It was not overgrown with greenery and the descent path was quite clear. I asked my co-pilot how he would decide to make his approach, how he would recon wind strength and control his descent, all the things a pilot must decide before attempting to drop below the tree tops.

As we circled I noticed some native people assembling and waving up to us from the area in front of the landing mat. Good, the tribe was still there. I told my co-pilot to make the descent. Steadily, at a slow forward pace, the helicopter approached rim of the clearing and began a slow rate of descent while still creeping forward. The final fifty feet or so of the descent was purely vertical above the mat. Any further move ahead and the rotor blades would strike a tree, with disastrous results.

"Steady," I murmured, as the new pilot tensed on the controls, "You're doing fine."

The four wheels touched down gently, and everyone relaxed as we reduced power to idling. I shouted down to the army bod to stay inside with the doors closed and reminded my co-pilot to keep the controls central as I climbed down. The small group of tribespeople gathered around, touching me and trying to shake my hand. Obviously they were glad to see me.

I walked back to the sliding cabin door and opened it. There was a loud gasp of awe as the little Jahai Negrito laddie stepped out into the open and ran to the arms of his parents. He was still dressed as he was in the hospital and had a little bag of goodies in his hand. To the superstitious aborigines this was like a miracle. The boy had not died! Here he was, dressed like the white men and obviously in good health. Here he was, a happy boy trying to tell his parents about the things he had seen after flying out of this clearing. But how could he do that? There were no words in their limited dialect to describe the traffic in Ipoh, ice cream, or any other thing he had experienced. It must have been baffling and indescribably difficult to convey. I had to interrupt to let them know that I had to depart again. I gave the boy a final embrace and shook his hand, waved to the people, and climbed up to the high cockpit of the helicopter. A nod to my co-pilot and he restored the rotor r.p.m. to operational status, then we climbed away. I took one final glance at the little motley congregation there below us.

Finding the new operational area was another lesson for my new pilot and we finally dropped our spare passenger, as well as some mail to a happy group of Kiwis now on duty there.

"Take us back to Ipoh now," I requested, sitting back to enjoy the view and musing on our little venture.

How would the boy settle back into the jungle after what he had seen and learned outside? Would he be happy? Would he possibly rise to become chief of his little group, or would he perhaps be considered a little mad and avoided? I wondered then, and have often wondered since. How long did he live in the jungle after we departed? What became of him? I only know the satisfaction of having been partly responsible for perhaps saving his life, but had I spoiled him forever by taking him from his native habitat into one as jumbled, noisy, and crazy as the full Malayan scenes of a large city? I guess we will never know the answers.

WHITE SILENCE

Snowflakes falling soundlessly on velvet air,

Cushioning the earth from man's rude stare,

An ermine mantle, each separate flake

Closely examined, reveals a mosaic

Of delicate crystals patterned like lace,

Each one different but equal in grace.

High mountain ranges all clad in white

Create a most magnificent sight.

How many snowflakes, each one unique,

Does it require to cover each peak?

This is no accident. This is design,

For man cannot equal this plan divine.

In such white silence bear witness I plead.

There's beauty around us, if only we heed.

187

FLYING FAMINE RELIEF IN ETHIOPIA

Much has been said and written about the famines in Ethiopia and other parts of Africa. Television viewers have had harrowing scenes of death and deprivation projected into their living rooms, followed by pleas for contributions to a worthy cause. In answer to these pleas a great many people throughout the free world responded with sympathy and an open-hearted generosity, almost without equal in recorded history. I saw firsthand how this world aid was helping the needy people of that drought-stricken land. Along with other volunteers I flew many missions of mercy, distributing the much-needed grains and medical aid to a starving and sick population in Ethiopia's most isolated, mountainous regions.

Although there was much evidence of gross wastage due to improper handling and poorly managed storage facilities, I witnessed the greater part of these donated supplies were indeed being distributed amongst those tribespeople most in need of help. However, we were only allowed in areas approved by the Marxist government of the time. Unfortunately there was also strong evidence that the black market was also flourishing. In the capital city of Addis Ababa and in smaller towns I saw many sacks of grain, each bearing the flag of some donor country, being sold openly in markets to buyers who were obviously not in dire need, this despite the printed logo on the sacks which stated in bold letters," A GIFT FROM THE PEOPLE OF____NOT FOR RESALE OR EXCHANGE." The names of Canada, USA, Great Britain, and France were prominent.

I had flown Bell helicopters close to the Ethiopian border previously while on contract to an oil company, so I was already familiar with the terrain and weather conditions. Over a period of years I had witnessed the creeping onslaught of the drought and had watched hitherto fertile grasslands die with appalling, deadly consequence to the animals, grain harvests, and to the people who strove to eke out a living there. My new contract however, would take me deep into the hinterland of Ethiopia and open new frontiers to me.

I was met at Addis Ababa airport by Don Wederfort, the area manager for our helicopter company. He was a well-seasoned campaigner in the Ethiopian theatre who had once been kidnapped by Eritrean Guerrillas. He quickly whisked me through the very involved Arrivals/Customs/Immigration formalities by waving special passes and visas in the faces of officialdom, but I was still subjected to currency checks and searched by over-zealous Customs officers who were interested in cameras, radios, and any other electronic equipment. I passed muster but I learned many such items found were usually confiscated on the spot, supposedly to be returned to their owners when departing the country…providing one could find the appropriate officials prior to boarding one's aircraft!

During our drive into the capital city along the only decently-paved stretch of highway in the entire country, Don filled me in on my duties. I would be flying a twin-engine Bell 212 helicopter (CG-ZII) in the high mountainous area of Ethiopia, about 100 miles to the north of Addis Ababa. The contract was with a religious group known as 100 Huntley Street. They would pay for the use of the helicopter and dictate where it would operate, as well as who would be cleared to fly in it. It was estimated that we would fly approximately eighty hours each month, carrying much needed supplies to the most remote mountain communities.

The next morning was spent obtaining airport passes and other documents which would allow me access to and from the ramp area. These papers all had to include passport-sized photgraphs, and a myriad of stamps. I also met a number of local authorities who would wield a high degree of control over our movements, whenever and wherever our flights were scheduled. I had expected to be flying at, or near the Sudan/Ethiopia border and had only brought lightweight clothing with me, so Don lent me his heavier woolen jacket. I filed a flight-plan to my first destination, Alem Ketema, about eighty kilometres to the north.

Grant Louden, the pilot I was relieving for a well-earned rest, accompanied me on this first flight and filled me in on the various procedures adopted to make the operation run smoothly. He pointed out useful landmarks and advised me on the best routes to follow whenever the weather showed signs of closing in. Addis Ababa sits more than 7,500 feet above sea level and is ringed by mountains lofting to over 10,000 feet. These make any blind approaches or departures rather hazardous during adverse weather conditions.

Alem Ketema (which we quickly shortened to A.K.) turned out to be a fair-sized village with a Baptist Mission established on its outskirts, close to a hard-packed dirt runway. This was long enough, and quite suitable for the ubiquitous De Havilland Twin Otter. These Canadian-built Jacks-of-all-work, owned and operated by such organisations as World Vision and Air Service, were crewed by Canadian and American volunteers whose jobs were to keep supplies rolling to all the remote landing strips scattered throughout the country. A.K. also sat astride a crude gravel road which wound tortuously all the way back to Addis Ababa, a grueling ride for the crews who handled huge Mercedes trucks laden with drums of aviation fuel, or sacks of grain from warehouses in the capital city.

Arriving at A.K. I was introduced to Val Hudilin, flight engineer for the base. He showed me around our accommodations, which consisted of several tin shacks erected over cement bases. These were reasonably furnished with a few chairs, a table, and single beds fitted with clean sheets, blankets, and a pillow. A propane gas stove enabled the crews to cook their own meals from supplies delivered weekly. Kerosene lanterns illuminated our quarters after dark, and a kerosene heater helped to take the chill out of the night air. Toilet facilities took the form of a screened hole in the ground next door to a crudely erected shower stall, where hot water was occasionally available. All in all, it was not too uncomfortable a place to operate from.

My helicopter was expected to serve several isolated communities from Alem Ketema, such as Gundo Meskel, Maranya, Rama, and Shil Afaf, on a daily basis. Two other settlements, Mehal Meda, and Rabel (slightly further east) needed only to be served once a month. I was not to enjoy the fleshpots of A.K. for long though because, after completing over thirty mercy flights, just two days later I was ordered to assist another church group operating under the name of "Food for the Hungry" based at a place called Sali, approximately 160 kilometres further north. It was situated on a gently-rounded plateau at an altitude of 10,000 feet above sea level.

My route took me over some of the most spectacular scenery I have ever had the good fortune to witness. Rolling ridges staggered across my flight path, rising in places to well over 10,000 feet. Many of the high points terminated in long plateaus which dropped off precipitously on all sides for several hundred feet. These housed peculiarly lonely and isolated communities, completely cut off from their neighbours by huge vertical chasms. Below sheer cliffs the land contours took on a gentler slope to form arable pastures and farming land, where livestock could graze for some distance before the earth plunged once more into shadowy gorges, cut like miniature Grand Canyons in the living rock. Occasional feathery waterfalls arced gracefully over these steep ledges, joining streams rushing to meet the muddy brown water of the so-called Blue Nile that paralleled our track.

About one hour after take-off the little camp at Sali hove into view. It looked quite forlorn beside a gravel-topped runway which had a distinctly visible downslope, dwarfed by several mountain peaks towering above. Two large canvas warehouses shaped like the old

Quonset Huts of WWII fame sat close to a pole carrying a dubiously useful, tattered windsock. A large fuel bladder sat next to the aircraft parking area, just clear of the sloping runway. About 100 yards away, a group of Tukuls (huts) were grouped inside a square enclosure that was surrounded by high thorned bushes. These were to be our sleeping and eating quarters. When I examined the huts I almost vomited, and felt like quitting right away. Compared to this, Alem Ketema was like Buckingham Palace! Walls of the Tukul were a combination of mud, grass, twigs, and sheep manure. The ground was its floor. Each boasted an ill-fitting door. Some roofs were thatched grasses and others had sheets of corrugated tin, a mixed blessing now the drought had ended.

These rustic dwellings were devoid of any furniture except for a crudely-made wood bed frame, strips of animal hide, or rope fibre, webbing to support an ancient mattress filled with questionable straw. Four grubby blankets and a hard, greasy pillow completed the inventory. Of course there was no electricity to provide light or heat, but we were denied even the comfort of a kerosene lantern to light our way. Stub of a candle sufficed to see us to our single beds after darkness had fallen, one to each tiny hut. The familiar native lavatory, screened for privacy and supporting its quota of flies, was simple to find…just follow your nose. There were no shower facilities here. Daily ablutions were limited to a quick sluice of cold rain water dashed over hands, face, and neck, from a water-filled 45 gallon fuel drum. It was so cold and, with no place to hang our clothes, most of us slept or at least retired to bed with all of them on. The mattress and blanket proved to be inhabited by hordes of voracious fleas, hungry for fresh blood and determined to keep us awake.

Central to this joyous oasis was the Mess, a slightly larger hut built of the same muddy ingredients, and very gloomy inside. Here we sat and awaited the whims of our chef, who served up completely unidentifiable and equally inedible meals. Seeing the meat bloody and covered with flies, hanging on the mud wall of the so-called kitchen, tended to take away my appetite. I skipped more meals than I ate and lost about 28 pounds during my tour there. Later on in my tour of duty, I found a supply of hard tack biscuits left behind by British troops in 1928. They were in sealed tins and quickly became my staple daily diet, shared with the children in the mountains.

My first task at Sali was to fly teams of doctors, nurses, and observer/interpreters to the more isolated settlements where the Twin Otters could not land. At each site the teams would interview gathered people, amassing all relevant information as to population density, livestock numbers, seed crops held, and the local mortality rate in census. Every child would be carefully examined, names, age, sex, height and weight registered. From previously compiled tables, doctors could then determine the degree of malnourishment in that district.

When it was decided that a chosen settlement required food aid, a scene of great activity would then blossom around the warehouse. Four huge cargo nets were spread on the ground at the helicopter pad, ready to be loaded with bags of assorted grains, milk powder, and medicine. Each bag weighing 110 pounds and bearing the flag of its donor country would be tallied against a check sheet before being added to the growing pile. As pilot, I had already calculated how much the helicopter could carry on each flight and instructed the warehouseman not to exceed that amount. This was always a compromise between distance involved in the round trip and minimum fuel requirements, considering weather and altitude. By keeping the fuel at minimum safety, I could increase the payload in the net and move the maximum amount of food possible on each trip. Since most of the flights were relatively short I had decided 350 pounds of fuel would safely do. This was good for about thirty minutes of flight time, and would allow me to carry 3,600 pounds in each net load.

Next procedure was to hover over the prepared load so a ground-handler could attach net to cargo hook on the helicopter belly before lift-off. I took a momentary pause in the hover to check power being used, that all temperatures and pressures were in the green, and with power enough to transition safely into forward flight we would be on our way. Maximum forward speed for a 212 carrying a heavy sling load is 80 knots and in most cases, with a dense load hanging motionless below centre of gravity, we could do this safely. At the drop off point I would bring the helicopter to a slightly high hover, and slowly descend until net gently touched ground before punching the net-release button on my Cyclic Stick. Checking that the load had actually fallen free of the hook, I would depart at maximum speed back to the heli-pad to land and throttle back to flight idle on power.

A mechanic would then do what we called a "hot refueling" and top the tank up to the specified 350lbs. This actually saved a lot of time and cut down wear and tear on the engine systems, caused by numerous starts. It was quite safe as long as the correct procedures were followed. An average of only four minutes on the landing pad saw us ready to pick up the next load and depart. When the fourth load had been delivered to the drop off, I would set down nearby and have the three empty nets tossed into the helicopter, ready to return to

Sali. In this fashion, the only time spent on the ground was for refueling or picking up empty nets.

At 3,600lbs per load, I was able to move about 50 metric tons on a good day, spending about eleven hours at the controls and logging about 9 hours of actual flying time for around thirty trips. It was very tiring and demanding work, but the sense of true accomplishment at the end of each day made it all worthwhile. I felt that I was actually contributing to a worthy cause, and could witness the effects on the people I was assisting.

Quite often we got a break because of inclement weather or occasional small snags in the helicopter equipment. At the start of the operation we had a number of long, very strong lanyards attached to the net at one end and the helicopter hook at the other. This kept loaded nets well below the belly and only one end ring was attached to the hook, speeding up each load attachment. In time, some of the lanyards frayed to the point of being unusable, while others simply disappeared, until eventually the person who was attaching loads had to crouch below the helicopter, attaching the four corner rings of each net directly to the cargo hook. I had to hover much lower over each pick-up load as a result of this, and again when dropping loads off. On one unforgettable trip this lack of lanyards created a problem I hadn't previously considered.

En route to a drop off point, I encountered very heavy rain and had to reduce forward speed in order to see ahead with safety. Bags of grain, exposed in the net, became saturated, increasing cargo weight. Consequently I had to increase power to maintain altitude, so fuel consumption went up. I considered jettisoning the precious load, or depositing it at some safe spot for a later pick up but was almost to my destination, so I pressed on. At the drop off point my load, now weighing much more than the original 3,400lbs, was too heavy for trying to hover. I trickled off the speed and allowed the machine to sink slowly until net touched ground. Slacking off tension on the net, I punched the hook-release button…and nothing happened! The load remained firmly attached to the hook. I tried the electric release and emergency foot, to no avail. Now I was in real trouble. I didn't have enough fuel to fly back to base at 80 knots. With no lanyard, I could not ease sideways or backwards to set down clear of the net load. A lanyard would have permitted this. I could not set the helicopter down atop the load, the 212 would simply have rolled over with disastrous results.

At this point, to add to my problems, the LOW FUEL warning light blinked red at me, indicating that I had roughly ten minutes flying time left before fuel starvation failed the engines. I waved my hand at one man standing by to indicate my problem and thankfully, he seemed to understand my predicament. He ducked low under my machine for what seemed like ages before I saw the net fall away in my underslung rearview mirror. I whizzed back to base as quickly as possible, in a sort of shallow dive attitude to increase my speed. Eight very anxious minutes later I gratefully ground the 212 skids onto the heli-pad, muttering a prayer of thanks. Two minutes or less to dead engines in that mountainous area was cutting it too close. Later on I found that the fellow doing hook up of the load had forced more than the usual four net rings onto the hook, which jammed it and prevented normal opening. I made sure all handlers were properly briefed after that scary episode, and prayed that extra lanyards would soon be found.

Our few trips back to the capital city were a real boon. The 212 helicopter had to go there for periodic servicing, which meant we were able to sleep in nice clean beds and use all the bathroom facilities available in our hotel room. To bathe and shower was bliss!

The long drought, experienced for so many years, had ended with a vengeance and the interior highlands were being subjected to extremely heavy downpours almost every day. Most afternoons huge cumulonimbus clouds roiled across the sky, piling in conflict against mountain peaks. Ear-splitting peals of thunder resounded across the valleys, and dazzling lightning bolts illuminated the skies in an almost constant display of Mother Nature's power. Precipitation usually took the form of torrential rain, often turning to severe hail storms with drops larger than marbles. They pounded the poor farmers' freshly-sprouting shoots of grain into useless pulp, covering the fields several inches deep in icy droplets.

These localised and vicious storms did not make my task any easier. Forward vision was often restricted to mere yards, the air extremely turbulent, and intended landing sites or safer flight routes were often enshrouded in hill fog. On occasion, these weather changes at mountain drop off points called for unusual and unorthodox departure techniques.

Quite often clouds sat on the plateaus like a top hat, creating fog conditions, assisted by the rotor downwash. If conditions proved too bad, of course I just shut down the two engines and waited for an improvement. But occasionally I could hover forward slowly, until the empty helicopter was poised right on edge of the sheer precipice. Aware that the cloud base was quite high in the valley below, I would transition blindly into forward flight until clear of the cliff side and drop the collective lever, putting the helicopter into autorotation. The high rate of descent, thus induced, would drop me clear of cloud within several seconds and my all-round vision would be quickly restored. This technique calls for a steady hand, instrument flight capability, and a good knowledge of local terrain.

When approaching a cloud-enshrouded drop off point with an underslung load on the hook, I would remain below the cloud until my main rotor tips were close to the visible cliff face and then inch up slowly until floor of the plateau was reached. Quite often there would be just enough space and visibility to slide forward over the escarpment and hover visually to the drop off point where I could release my load. If no opening appeared, I would just turn away from the cliff face and head quickly forward and downward to get below cloud as fast as possible. Of course we always checked cloud base and altimetre readings constantly before venturing to use this technique.

Each time the helicopter appeared with a net load of supplies people assembled near the pad regardless of weather. There they stood, or crouched, with their bare feet squelching in thick gooey mud, their clothing in tatters, and their faces seemingly aloof or expressionless. How could they know who we were, or where we came from? They knew nothing of Canada, or America, or any other place outside of their parochial mentality. Yet when I had occasion to take a welcome break from the cockpit and walk around, they stood, smiled, and made a sort of polite bow to me to acknowledge my contribution. If I shook hands with one, I had to shake hands with all present. I would, on these occasions share my hoard of hard tack biscuits with the children present. "Bis-i-kit" became a well-known shout and probably the only word in English these youngsters learned. When each family received their allotment of grain and milk powder, they headed home with the bags over their shoulders or slung over a donkey's back. I often saw people wearing empty sacks like a monk's cowl over their heads and shoulders. One such family had the proud red maple leaf of Canada boldly emblazoned across their shoulders, and that sight made me feel good about my adopted country.

The tragedy was, these poor souls were completely dependent upon government whim for their relief supplies. In one such settlement about 10,000 feet above sea level the Baptist Minister from America, in his dual role as a doctor, told me that the local communist official was using the food/grain distribution meetings to brainwash the inhabitants. Before getting their free rations, each family had to attend indoctrination lectures extolling the virtues of Communism. They were further led to believe that all the life-giving supplies being flown in were a gift from their benevolent government. In yet another small community, the population was duped into paying a "Head Tax" related to the size of the family, before their "free" supplies were handed out.

Our area of operations was always very clearly defined by the authorities and all too often some bumptious official hanging around the helicopter pad would insist that he had to accompany us to our destination. Although regulations forbade passengers when carrying sling loads, we were often forced to comply and had to reduce our payload accordingly to avoid exceeding our maximum "all up" weight. Certain areas north of Sali and south of Addis Ababa were forbidden to us, and no food supplies were permitted for the Eritrean area, where rebels opposed the ruling Marxist government. The unfortunate souls who lived in those areas (who were not all rebels) were left to their own devices, to starve, or to emerge and be captured or killed by the Federal Army. This was a pitiful situation which should never have been tolerated by anyone.

FAMINE IN THE LAND

Famine reaches out a bony hand,

and plucks flesh from wasted human frames.

Flies suck the last vestiges of moisture

from eyes, nose, and mouth as they

plant seeds of painful death.

Children too apathetic to even wave insects away

stare at the world with large sad eyes,

while malaria, bilharzia, and fevers lay them low.

How short and unhappy is their existence.

Fate knows no crueler way of crushing hope

than to watch children die needlessly

in an uncaring world.

Desert creeps inexorably over land,

swallowing pastures and trees in smothering sand.

Rivers dwindle and die whilst oases run dry,

water disappears to add to the misery

of people who have no hope for any future.

BOMBAY MERCY FLIGHT

What am I doing up here at five-thousand feet above the Indian Ocean, flying away from the coastline in the middle of a thunderstorm? I asked myself this question as torrential rain swept over my helicopter's windshield in a veritable bow wave, and several towering cumulonimbus clouds glowed ominously close on my radar screen. The answer was quite simple. There had been an explosion on one of the many oil rigs out at sea; several men had been burned and badly injured. Two men needed to be rushed to hospital as quickly as possible, so the helicopter had been called into emergency action. But I am getting ahead of myself. Let's go back to the beginning.

Employed as pilot by Okanagan Helicopters Ltd, I was based in India during most of the year 1979 and living in a Holiday Inn in Juhu, a suburb of Bombay. Our helicopter was a twin turbine-engine Bell 212, with the registration of G.O.K.L. (Golf, Oscar, Kilo, Lima) for call signals. We were contracted to fly men and materials to and from the land base and a number of oil rigs and drilling ships, about 150 miles offshore. During the dry season, when visibility was excellent, day or night flights presented little or no problem for the pilots and engineers crewing the helicopter, but things were vastly different during the wet Monsoon season. This usually started around early June and continued for several months of extremely cloudy, rainy, and stormy weather. It was then the pilots really earned their flight pay. Under I.F.R. (Instrument Flight Rules) we had to be qualified. That is to say, we had to be capable of conducting all flights with reference only to the aircraft's flight instruments, using radio, radar, and navigational aids for assistance without benefit of outside visual references. The Department of Transport Air Regulations also decreed there must be a minimum crew of at least two qualified I.F.R. pilots at the controls for all flights, and they must be familiar with local blind-flying procedures.

Since we were living in comfortable hotel quarters during our two monthly stints in Bombay, many of the Okanagan crews had their spouses join them. My wife Bunnie was there with me during this working period. Enjoying a rare day off, Bunnie and I had gone shopping in a Bombay market despite the rain and had returned to the hotel in early evening. We sat to enjoy a meal in the well-frequented dining room where our favourite waitress danced attendance upon us. She was a beautiful young lady named Sarita, and she knew all of our helicopter people by first names. I commented on the fact that there seemed to be no other Okanagan crews around and Sarita told us that they had all been invited to an Engineer Rep's house. He was the Rolls Royce specialist and had just moved into a sumptuous new apartment, so decided to have a party. Not being present at the time, we had missed the invitation. I was scheduled as Duty Pilot for the next morning so Bunnie and I went upstairs to our room.

We had started to get ready for bed when the telephone jangled. I answered and heard a familiar voice, asking me who the Duty Standby Pilots were. I told this senior Oil and Natural Gas contractor the names and he replied that he'd tried to locate them, and failed. He then told me of an explosion on one of the rigs. Two men had been badly injured and needed to be flown to hospital.

He stated the urgency in getting GOKL into the air as soon as possible. I assured him I would see to this, and he broke the connection.. Knowing that the standby pilots were not in the hotel (which was breaking the rules) I decided that there was only one answer. I would have to bend the rules a little bit myself. I dressed quickly in my flight suit and grabbed my helmet before dashing downstairs to catch a waiting taxi.

"Take me to Juhu Military Airport, *jildi*!" and off we went in the rain. At the Juhu guard house I had a soldier accompany my taxi across the airfield to where the Bell 212 sat. By the time I had finished walk around, pre-flight inspection, untied the rotor blades and unplugged the engine intakes I was saturated, but commenced to start both engines, switching on the radios and radar equipment. In a 212, the Captain sits in the right hand seat and the co-pilot on the left attends to radar, radio, and navigational gear etc.

In about two minutes I was asking the control tower at the main Bombay airport (Juhu military was closed down after dark) for an emergency clearance take-off at Juhu, outbound for the rigs. I was given clearance for take-off and instructed to remain at, or below, five hundred feet until five miles off the coast, outbound. As soon as I crossed the shoreline at the end of the runway I was immediately on instruments and climbing to five-hundred

TIME JUST FLEW

feet in extremely bumpy air conditions. Five miles out I was further cleared to 5,000 feet and ordered to follow the westerly radial of 270 degrees. This was a compass heading I had to fly for identification on the Air Traffic radar. Level at 5,000 feet, I set cruising power and trimmed the auto pilot to hold height and heading. I had to loosen my safety harness a bit so I could sort of slide over sideways to see the radar screen. As if I didn't already know, the screen showed several active thunderheads around me. I could see the intermittent lightning flickering inside them already, and they were directly ahead of me.

"Not going to be a smooth flight."

It was almost like flying underwater. I was jolted up and down, bow waves on the front and side windshields. No point in trying the wipers here. I was cocooned in a noisy bubble listening to various aircraft radios chattering but I was all alone, no one to speak to for about an hour, when I changed frequency to that of the rig I was heading for. I reported my height, 5,000 feet, and distance, 20 miles from the oil rig at this required point. I was then cleared to descend, using their radio beacon as a directional guide for me to hone onto. The descent was also very bumpy and, of course by this time I was flying manually, holding all controls on my own. Descending very slowly, about three-hundred feet per minute in complete darkness, I stayed on instruments and asked the rig operator to confirm his altimeter pressure setting. I aligned GOKL's altimeter to his specifications. At two-hundred feet above sea level I was still in cloud and the rig showed on radar as five miles straight ahead.

Talking almost constantly to the rig radioman now, I asked for all lights on the rig to be switched on and then descended to one-hundred feet. I could just make out the waves and flying spray if I glanced outside. Switching on landing lights, I pointed the beam ahead and downward about forty-five degrees. I reduced forward speed from an indicated 120mph, to 45mph. With the headwind of 25mph, my rate of travel forward over the water was now only about 20mph. To attempt to fly any lower than 100 feet in height would be tempting fate and a watery demise, but there was also a danger of possible collision with the mast of any of the numerous small boats anchored around the rig. I couldn't take my eyes off the water now and kept staring ahead through the wipers, searching. Ah! At last I could see a glow of brilliant lights as I neared the steely long legs of the rig. I brought the helicopter to a hover, the helipad now one-hundred feet above me. I eased up vertically until I saw the large H of the pad appear with a rope net secured across it. This would prevent the metal skids of the chopper from sliding on bare steel plating. Gratefully I reduced power, touched down, and gave the thumbs up signal for the stretcher-bearers to approach the helicopter's sliding door. The boss man of the rig approached from the left and opened the co-pilot door. The expression on his face said it all.

Where the hell is your co-pilot?

I shouted over,

"I will explain that tomorrow morning when I'm officially reporting for duty!"

He nodded okay.

Two stretchers were now securely tied to the metal floor behind me and a co-worker had been assigned to travel with me as helper to keep an eye on the casualties during our return flight. When he was strapped-in I gave the nod to everyone that we were about to lift off and opened the twin throttles. A slight pull up on the collective lever and we were airborne again. Pushing the nose down, I climbed forward into wind and immediately went onto instruments, the rig sliding behind us. Back up to five-thousand feet and a reverse course of 090 to head for the shore once more. There was no conversation between us, and no helmet for the helper. He looked rather scared.

The tailwind shortened our return flight somewhat and when I picked up the shoreline on radar at ten miles, I once again descended to five-hundred feet, slowed down my speed, and called Bombay International Air Traffic Control. Officially Juhu military was closed, but I told the main terminal that I was being met there and the lights would be on for me. At five miles offshore I was cleared to head into the darkness of Juhu, where I was able to pick up the lights of hotels lining the beach. I flew sideways until I spotted the gap where the runway met the sandy shore. An ambulance was waiting by a hangar near the helipad with his headlights on high beam, so landing was no problem. Once the stretchers were removed and the ambulance was on its way, I shut down and secured GOKL. The trek across the airfield was another wet one. Finally I got a taxi at the main gate and was back in our hotel room shortly after midnight, soaked but safe.

Next morning I tore a strip off the slightly hungover base manager and his co-pilot. They had been the standby crew for last night. I told them of the Cas-Evacs and mentioned that, if these two injured men had not been airlifted to hospital, they could have died. Both pilots would have been in serious trouble and Okanagan Helicopters might have lost a very profitable contract. The two injured men, I'm glad to say, recovered. The official log book of the flight was cooked to show that two pilots had been aboard GOKL, so all was well.

OLD BOMBER BASE REVISITED

Deserted, abandoned, an airfield spans the lonely
heath,

Unkempt broken runways sprout their share of grass
and weeds.

Bare dispersal pans of circular concrete sit, now
empty,

Lacking the black silhouettes of bombers

Which used to squat, etched against the darkening
sky.

Empty pre-fab huts with broken, glassless windows

Gaze sightlessly out at overgrown hedgerows,

And seem to echo back the voices and laughter of
youths

Who, in blue, used to ride the night skies to
destruction and death.

Wind sighs in lonely desolation, as if recalling

The vibrant roar of countless Merlins,

Coughing puffs of blue smoke to be whisked away

In the swirling prop wash of so many Lancs

Ponderously thundering into the clouds, and when
massing,

Made the very earth tremble with their passing.

Where ground crews kept tally of every departure,

And murmured a prayer for each aerial charger.

Throughout long nights, their vigils maintained,

Till in the grey dawn, their visages strained,

They counted the losses.

A strange way of life, to protect a way of living.

Sacrifice demanded, and the ultimate too oft' given.

A lonely figure walked to where a runway ended,

Thoughts deep in the past, as his spirit blended.

217

Seeing this airfield as he once knew it,

Remembering well, he'd flown, and lived through it.

Exorcising ghosts, he roamed o'er the acres,

Recalling faces, nicknames, givers and takers,

Pilots and Navs, Flight Engineers too,

Wireless Op's, Gunners, from each motley crew.

But the vision all vanished. The noises all dimmed,

Till all that remained was the sigh of the wind,

A creaking window, the rustle of grass,

He returned to the present, and bade adieu to the past.

About the Author

Jim McCorkle was born in Glasgow, Scotland, 1921. He joined the Royal Air Force as a young man, to learn how to fly, and to serve his country in WWII. He did further training and received his wings at Pensacola Naval Air Base in Florida, September 10th 1942. "Mac" was decorated by the Queen, and received many other accolades in his twenty-five years of military service, all over the globe, especially in Southeast Asia. He met his wife "Bunnie" during the war, celebrating many years together before her death in May of 2012.

After his R.A.F. service, Jim and his family immigrated to Toronto, Canada. He became a bush helicopter pilot, away from his family for nine months of the year. He tried to fly a desk job with the Canadian Air Ministry (re-writing the manual for civilian pilots while he was there) but returned to the freedom of bush-flying after one year.

Besides flying in B.C.'s Rocky Mountains and the Canadian Arctic, he signed-up for jobs that took him to Africa, flying famine-relief in the Sudan, and India, flying men and equipment on and off oil rigs. When possible, Bunnie accompanied him to other countries. They both loved to travel, especially to warm countries, or often back to England.

At 93, Jim McCorkle still takes part in "The Memory Project" speaking to school children about his love of flying, and the honour of serving one's country. He stays active with daily walks, still loves to read, and keeps in touch with family and friends worldwide on his computer.

Made in the USA
San Bernardino, CA
25 March 2015